This book may be returned to any Wiltshire
Library. To renew this book, phone or visit our
website: www.wiltshire.gov.uk/libraries

Wiltshire Council
Where everybody matters

LM6.108.5 (CH 2017)

# Sacred Journey

## STONE CIRCLES & PAGAN PATHS

## SALLY GRIFFYN

KYLE CATHIE LIMITED

First published in Great Britain in 2000
by
Kyle Cathie Limited
122 Arlington Road
London NW1 7HP

ISBN 1 85626 364 9

Edited by Caroline Taggart
Designed by Kit Johnson
Map by ML Design

A Cataloguing in Publication record
for this title is available from the
British Library.
Printed and bound in Singapore by
Kyodo Printing Co.

*Half-title: Maze carving near
St Nectan's Kieve
Title page main picture: The Ring of
Broadger, Orkney; small picture:
offering left at St Patrick's Well,
The Burren*

*ACKNOWLEDGEMENTS  I would like to thank the many
friends who have supported me during this journey, in
particular Jan Arnold, Witch and cosmic sister, who travelled
with me to some of the sites in Ireland and England; my
friends Saira and Mikey Dawson, who know the ley of the
land on the Burren; Ulric Whyte, for his encouragement and
yogic companionship along the way as well as his excellent
web design; Kathy Srabian, for her wise advice; Michael
Parker, for his enthusiasm at the Ring of Broadger and the
night-time visits; Chuck Pettis, who sat with me in the stones
of Callanish and sent me info on publishers; Ambrose and
Mawkin for being there; Margaret and Ron Curtis for their
invaluable knowledge of Callanish; my father Tom Hutchinson
for getting me my first camera and enthusing about my
photography; Hugh the Witch, who taught me about
language; and Gwynn, who lifted the veil on Slea Head and
thereafter nothing has ever been the same. The Witches and
Pagans who taught me the old ways are the hidden ones who
mapped out the sacred sites in this book.*

*I would also like to thank my editor Caroline Taggart for all her
inspiration and hard work, Kyle Cathie for believing in the
project, Mark Buckingham for his amazing cover artwork and
Kit and David Johnson for their wonderful interior layout.*

For information about workshops, talks, retreats and
tours, write to Sally Griffyn care of the publishers, Kyle
Cathie Limited, 122 Arlington Road, London NW1 7HP.
All correspondence will be passed on to her. Or contact
www.sacredjourneys.co.uk;
enquiries@sacredjourneys.co.uk

# Contents

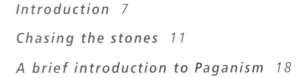

Introduction *7*

Chasing the stones *11*

A brief introduction to Paganism *18*

**STONE CIRCLES, RITUAL SPACE** *23*

**THE EARTH GODDESS AND
THE GODDESS OF THE LAND** *51*

**MAID, MOTHER AND CRONE** *65*

**MAGIC FROM THE SEA** *79*

**SACRED WELLS, HEALING WATER** *97*

**SACRED LANDSCAPE:
THE GODDESS AT AVEBURY** *123*

**THE ISLE OF AVALON** *137*

**PASSAGE TOMBS
AND CHAMBER TOMBS** *149*

Afterword *165*

How to get there *169*

Glossary *172*

References & further reading *174*

Index *176*

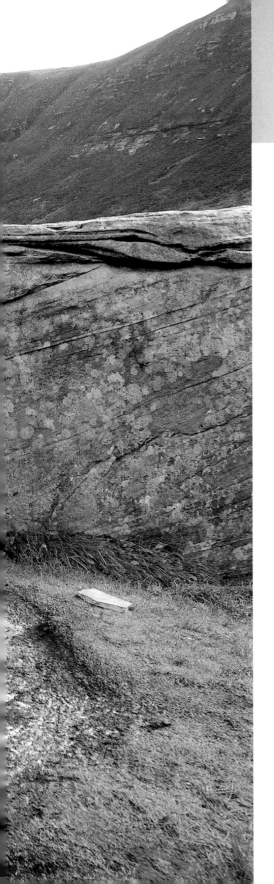

# 1NTRODUCTION

On the eve of a new millennium my commitment to writing this book came as part of a journey I had undergone in visiting sacred sites. There is no doubt that there is a raising of spiritual awareness as we move into the Aquarian Age. This, coupled with the desire expressed by many women to find an image of Goddess in an array of male-focused religions, led me to seek out ancient spiritual centres that are surrounded by female mystery.

Sometimes folklore holds a clue to where these centres are, as in the case of the Hag's Chair at Loughcrew in Eire. I followed the stories and asked the locals. Some of the sites I visited have no name, no stories, and merely stand as relics of the old religion. They are silent and can be accessed only by touch. My earlier work as an archaeologist was the key to those silent spaces. I learned how to look at the land and when no obvious answers were to be found, learned how to ask the questions.

The body of lore that is contained in Pagan practice enabled me to ask with my hands, with my intuition and with ritual. The stones that speak in whispers come alive when they are revisited with ritual. In this book and on this journey I encourage archaeologists and those interested in folklore to change their approach to sacred sites: to cast off inhibitions about being seen to be wacky and to use ritual at these places. Ritual need not be elaborate – it may involve no more than entering a sacred space, lighting a candle and meditating. If you have never written poetry, then know that these places are rich in poetic expression. If you have never chanted, then bring music into the earth tombs and feel them echo with timeless energy.

**THE DWARFIE STANE** *Hoy, Orkney (previous page). Legend speaks of giants moving the stone into place. The interior has two chambers carved out of massive rock. Opposite, washing Witches' sacred tools in the Blood Spring, Chalice Well, Glastonbury.*

It is worth mentioning that I am a Witch. This came later on my sacred journey, for it took time to find expression for my spirit path. I had always been interested in the tarot, meditation, alternative medicine, women's spirituality and folklore. In early 1990 I attended Vivianne and Chris Crowley's Wiccan workshops in London. There I found myself drawn to a path that held the ancient knowledge of the land. For though I had studied some of the Eastern paths and fully recognise their teaching, it was important for me to work with the land where I lived. I was initiated and began the training. As part of my work as a photographer I was commissioned to photograph Doreen Valiente, a Witch whose wealth of information on British magic and folklore led me to study folk practices and peasant beliefs. We became dear friends and when I started my own coven she came to work with me, thus spanning the generations. At the time she was 77 and I was 37, so there were 40 years of difference in approach.

Sacred sites have held ritual space for generations and continue to be portals through the veil. The rites of passage that accompany our journey on this planet, in this life, have been lost and need to be reinvented. I have had the opportunity to study with one of the Elders of the Craft who taught me to recreate the rituals that have been lost and not to be afraid of changing and reinventing these lost treasures. Our ancestors left their energy at these places and we have the opportunity to join our energy with theirs over the aeons. In this book I do not attempt to cover all the sacred sites of the British Isles, for there are many. Instead I look at some of the more remote sites and/or good examples of Neolithic ritual sites. I include ideas for you to try, and my own personal rituals. This is an invitation to all on a spiritual path to take my hand and to come … to dance, to feast, to sing and make love in the stones.

After I wrote this, but while I was still working on the book, Doreen Valiente died. She is considered by many to be the mother of modern Witchcraft or

Wicca and wrote much of the *Witches' Book of Shadows*, a handwritten work that is passed on from initiate to initiate. Doreen was a member of Gerald Gardner's original coven in the 1950s and was courageous in her efforts to make women in the Craft a force to be reckoned with. She was the author I first turned to when studying the ideas of Wicca and found her to be a wealth of information. When she attended my Craft of the Wise workshops, she observed that open workshops about Wicca would have been unthinkable in her day. The attitude to Witchcraft was and still is so negative. She was amazed that some of my workshops were in churches, though they were under the guise of Celtic and Pagan history.

*I DEDICATE this book to Ida Griffin Coyne, my beloved grandmother, who taught me to look at the sky.*
*I dedicate this book to the many unnamed Witches who hold the secrets of the land.*

We spent a few years together and went to some of the sacred sites in the south of England. Glastonbury was her spiritual centre and as she was dying and we meditated together, it was to Glastonbury that she went in her thoughts. Turning to me, the last time I saw her, she smiled and said, 'Do you know, I was just at the Tor?' The pull of the ancient sacred sites was so great that it was to the heart centre of Britain that she went on this last stage of her journey.

I include her in my thoughts throughout this book. I could not have known that the passages that I wrote on my journey to the stones paralleled her journey to the afterlife. She walked the path with me as she now walks in the Summer Lands.

*Merry meet and merry part*
*and merry meet again.*

*Blessings*

SALLY GRIFFYN

*THE STONES OF BRITAIN,* like the stones worn around those ancient necks, are reminders of an age-old religion. We attempt to step into ancestral shoes but there is little science to tell us how it was. We just know that women and men walked upon the same land that we walk upon today. Their footprints have faded much as ours will, but the stones are the finest examples of endurance that we have. They stand as monuments to a great culture, great art and complexity, great survival and wisdom. We can access them only through our hands and the resonances that continue to make up a body of lore and of people who feel extraordinary things in their midst. They seem to have strange weather patterns associated with them. They seem to be placed in lines and in places that sit easily in the landscape – somehow it all makes sense in contemporary terms even though the weather, the temperatures, the earth shifts, have changed.

## *AN UNNAMED STONE ALIGNMENT*

*(main picture) in the Ring of Kerry, Eire. Inset, a vast stone at Sainte-en-Tregastel, Brittany, decorated with interlacing Ss. Experts disagree whether it is a fertility symbol or a signpost indicating a nearby burial site.*

# *CHASING THE STONES*

The journey began when at the age of nine I walked into my grandmother's garden, looked up at the night sky and realised that I was connected to all that was. There, under the night sky, I felt the longings of the wild, the call of the Witch, the indescribable pull of the priestess. I stood alone. I was truly free. This was the deepest understanding of life that I had had. Nature was magic and I was changed.

It was my beloved Grandmother Ida who told me of the Witches. She told me that they ran naked on the moors, and that there was a White Witch in Sheffield who treated people with milk and was good and healing. My grandmother spoke of reincarnation and how she felt this was true; she also said that she saw things before they happened. Ida was my first spiritual teacher. It came from the blood and from time, and she was my great friend.

I remember going to Stonehenge with my parents. This was a time before fencing and barricades and men in uniform guarding the stones from the hippies. I wandered among the stones and they had a deep impact on my dreams. I felt time running through the stones. I felt the awe of the ancient. I was hooked.

Many years later, just out of university, I answered an advertisement for an archaeologist. I had no idea that we would be specialising in Neolithic ritual sites. We visited many sites and learned dowsing from our leader. We stood together in close circles in the inner sanctum of West Kennet barrow tomb and felt exhilarated. I had been following a strongly conscious magical path since the night of the stars and this gave me the opportunity to use some of it in the sites. I had the pleasure of seeing Chanctonbury Ring on the night of the 1987 hurricane; we climbed among the huge torn-up trees that had been at the centre of folklore about Witches. Although the ferocity of the storm had passed, the atmosphere that pervaded the ring was remarkable. I felt the pull of the ancestors. It involved emotions that I had come to recognise as part of the side of me that spoke of healers and priestesses and of the power of Nature.

### THE GREAT CIRCLE AT CALLANISH

*of all the sacred sites I have visited, I think the stones at Callanish are the most beautiful. Their twisted forms lean into one another as if they were communing.*

By this time I was well on the path. I was meeting locals and people who could teach me about crystals and sacred sites: they spoke of lone hills at night, of gypsies and travellers, of placing crystals on the body to heal, of reading tarot at sacred sites, of dowsing for the right crystal for the right price and the right use. I was gathering my store of this land. I was drumming and reading and visiting ancient sites. Later I loved someone who took me to sites and we ate mushrooms picked fresh from the land and made love under the full moon.

What I learned about the Earth was that it healed things in me that no words could console. It relieved me of human concerns and took me to a place, a small place, where my problems seemed insignificant in the face of such majesty. The moon, so beautiful and universal, must have been the first image of perfect symmetry that the ancestors gazed upon. I knew that the moon charged up the sacred sites and that you could feel her running through the stones like a cool heat. I also found out that the sites were still regularly used. I found necklaces of holed stones hanging in a tree at Chanctonbury Ring and knew that I stood on the site of a recent ritual. These holed stones speak of a time when tools, houses, utensils were all created from stone. The imagination at that time made leaps that we simply discard as primitive technology. But they were dealing with first-order creation – making something out of nothing.

When we enter the stones we enter a sacred place. The circles came at a time when there were no examples of symmetry except the moon at the full and

the sun at the rising or the setting. This is why stone circles are so fascinating. They are based on balance. Each stone stands beside and across from its neighbour, allowing all angles to be seen. Ritual and communication across space is made easy in stone circles no matter how large they are. Fires were lit among the stones and crystals buried under them. Human bones were placed beneath them; some even have bodies buried at their feet. In one case an inner ear bone, invisible to the naked eye and contained within a hollow inside the head, was buried under a stone. If this is a reference to sound, it is striking that there are rumours that the stones hum and were raised by sound.

These are the things I have heard on my path.

During one of my digs in the Hebrides, at the Summer Solstice, I took the day off and walked for miles with another archaeologist. It was on that day that for the first time I clearly saw ley lines in the landscape. They shimmered in the sun and I had an insight into the placing of stones and stone circles. The land in the Hebrides is uninterrupted by habitation and it is easy to see how the ancestors may have viewed it. I was travelling along the ley lines and realised why the journey seemed so easy – it was as if the lines were feeding me energy. That was when I realised that the lay of the land is really the *ley* of the land. The ley lines are the old straight tracks.

What is often forgotten is that the placing of stone circles and standing stones was not based merely on sight. 5000 years ago most of the landscape was covered with trees. What we talk about as ley lines, people often misconstrue as being visual patterns. They are indeed visible, but more often they are felt. It is as if a circle sits in the landscape in the most balanced manner, taking all attributes of the surrounding area into account, and then focuses the energy centre. When dowsers say a site has lost its energy, they mean that the land has changed, the ley has changed – not surprisingly, as this planet we live on is an evolving being. Mountains are thrown up and land breaks away. When we say a stone circle fits the land we really mean 'within the landscape', for we are not seeing it the way our ancestors saw it. What I saw on that Summer Solstice was a connection from one site to the next that was quite incredible considering that when they were built one would literally not have been able to see the next site. Nevertheless, one could feel where it was likely to be.

It was also apparent that dwellings were separate from ritual sites. Not much has changed in the way we keep our spiritual practices away from the home, in temples and meditation chambers. That is, unless we are taking our ritual into the home.

I had in fact been taking ritual into my home for many years, for there were no modern temples dedicated to the Goddess. It was because I wanted to work at pre-existing ritual sites that I took ritual back to the ancestral temples. It meant I had to be that much more dedicated. The Druids, much to their credit, have taken their cause to the European Court of Human Rights, protesting that they

are restricted from using their ancient temples. I, on the other hand, went underground, using stone circles and standing stones as a way of increasing my understanding of the ancient ways. I visited circles in the night, early morning and during the winter when they were less of a tourist haven. I had remarkable results. What could be construed as vivid imaginings have in part been backed up by others who visit the same sites in this way. I have felt stones giving and receiving. I have envisioned rituals with central themes that I had not explored myself. I have received comfort, information and well-being, and I have seen weather patterns that were truly phenomenal. I have grieved at the stones when loved ones moved over to the Summer Lands. I have made passionate love among the stones. I have lit fires and I have seen birds and animals gather with me at the stones. Once in Ireland – at Slea Head in Dingle – I heard the land speak as

I leaned against a stone that was not even marked as a sacred site. I was with my lover and did not know at the time that the stone that harboured us from the sea mist and driving rain was an ancient ritual site. The energy compelled us to lean against it and we were transformed by it. The Kundalini energy of sacred sex sites rises up from the land and warms the stones.

These places cannot be taken as a one-off experience. They can, of course, be visited once, but there is something important about the concept of pilgrimage and returning to the place of the ancestors again and again. I have noticed that every time I return, something new and transformational happens. Consciousness slips into dreamtime, where time is stretched or shortened, at sacred sites. Sometimes you believe that you have been there twenty minutes and you find that it has been three hours. This kind of distortion is referred to in folklore as faery time. Often in mythology a visitor is met by a stranger, an enigmatic being who asks them to visit their home inside a barrow or hill. The stranger seems human yet unfamiliar. Described as stunningly beautiful or incredibly ugly, short and dark with green eyes and seductive, they take many forms, including that of an animal. Sometimes the meeting takes place at a crossroads, at a standing stone or stone marker or in a circle of stones. The visitor feels compelled to accept the invitation, but when he or she comes out again seven years have elapsed.

*GALLARUS ORATORY* on the Dingle Peninsula, Éire. Created without mortar in the 7th century, this meditation chamber remains watertight. It is an early Christian cell in an area where Christian and Pagan symbols frequently exist side by side.

A hidden body of lore brings together the mythical and the historical, in dreamtime. Though we have no first-hand account of what went on at sacred sites, many myths and legends refer to the stone circles, standing stones, healing wells and barrow tombs as places 'between the worlds'. But the only way truly to know the stones is to be with them and to feel them.

If I can be that stranger on your path I would like to take you to the stones and let you in on a few secrets. I have become entranced by these temples to the sky. Like others, I am a seeker on the path. The journey to the stones is a reflection of the inner journey that one takes when partaking in ritual. Sacred

sites are places full of mystery. They are maps of the Earth's energy centres and reflect the emotional response of the visitor in the landscape. Some are happy and joyful, while others serve a different purpose. They are places of healing that discharge negative energy and fill the heart with the courage of a warrior. This book is an invitation to walk with a stranger and to chase the stones.

Although all the sites I visited in the course of writing this book are in the British Isles (including Eire), signs of Mother Goddess religions are found the world over. The same pre-Celtic themes can be traced through much of northern Europe. The Gundestrup Cauldron, discovered in Gundestrup in Denmark, shows the snake as a symbol of sexuality which is reflected in the processional avenues of Avebury. In the Teutoberg district of Germany, on the ancient rocky outcrop of Externsteine, are the remains of a Pagan temple whose window-like opening is precisely aligned so that sun shines through it at dawn on the Summer Solstice and the moon appears framed by it when it reaches the northern extreme of its orbit once every 18 years. The standing stone alignments of Carnac in north-western France form the greatest astronomy temple in Europe and are known to predate the Pyramids, Knossos or Stonehenge. France also abounds in holy wells, now mainly associated with Catholic churches but based on springs that have sacred connections much older than Christianity.

*TWO VOLUPTUOUS FIGURES of the goddess as a symbol of fertility. The Austrian Venus of Willendorf (bottom) is 30,000 years old; the other, from a 5000-year-old Neolithic flint mine, has the same emphasis on abundance and bounty.*

The Mother Goddess traditions of the American peoples and of Hawaii spring from a different culture, but allow American readers to visit sacred sites closer to home. The rock formations of Sedona, Arizona, are described as vortexes of the energy of the Earth. In the eastern and mid-western United States, chambered tombs are covered with mounds not dissimilar to the pre-Celtic long barrows. Many were enclosed in 'sacred circles', though these were massive earthworks rather than standing stones. The Adena people created effigy mounds, of which the most famous is the Serpent Mound in Ohio. We can only assume that this was a form of totem, but it is tempting to assume

*EXTERNSTEINE* is the sacred heartland of Germany. The rocks contain a mysterious series of caves that certainly predate the medieval monks known to have lived there.

that it also had sexual connotations. The worship of a fertility-oriented Mother Goddess is most evident in Hawaii, where sites dedicated to the goddess Pele can be found.

Whether visiting sacred sites close to home, or travelling long distances, be aware that the journey itself is as important as the ritual or the meditation that you will perform at the site, and travel with intent. For, as the great 20th-century occultist Aleister Crowley said, 'Every intentional act is a magical act.'

# A BRIEF INTRODUCTION TO PAGANISM

Within Wiccan lore there are teachings that are spoken as part of a process of invoking God and Goddess energies into a circle. Invoking, or 'aspecting' as it is called in the United States, is both a calling to and a calling in of the energy of that season. The spoken word is important in ritual at certain times and it is within the most famous 'charge' that the tenets of Wise Craft are found. The Charge of the Goddess was written in this form by Doreen Valiente, though the Pagan sources are far older. She revised this version in the 1950s and introduced it into the *Witches' Book of Shadows*.

*Whenever you have need of me, once in the month and better it be when the Moon is full, then ye shall assemble in some secret place and adore the spirit of me who am the Queen of all Witcheries...and ye shall be free from slavery and as a sign that ye be really free, ye shall be naked in your rites...and ye shall dance, sing, feast, make music and love, all in my praise...for mine is the ecstasy of the spirit; and mine also is joy on Earth, for my law is love unto all beings...for behold, all acts of love and pleasure are my rituals.* [1]

The Charge gives some indication of the ideals of Witchcraft and the free expression involved. Love is the highest ideal; a fear-based religion is actively against Pagan ethics. This goes some way towards explaining why modern Pagans feel alienated by some of the male-focused religions of the world. Women are honoured in Paganism. The burning times – a period of several hundred years when Witches were frequently burned at the stake or hanged – are a reflection of how fearful Christianity was of women in power.

Of course nothing like Amnesty International existed then, but some of the issues of women's freedom that it is addressing in certain eastern countries are the same as those in the burning times. In hard-line Muslim tradition it is still the right of the father and brothers to kill a woman who sleeps with a man other than her husband. The image of the promiscuous woman, like the concept of the Witch as seductress, is part of the misogynist culture that emanates from male-focused religions. The exact historical framework of the 'burning times' is hard to assess, but 'witch hunting' of one kind or another was common from the time of the Crusades to the late 18th century. The misdemeanours that led to death had many names: heresy, treason, Witchcraft, to name a few. In England the Witchcraft Laws were not repealed until 1951, when they became the False Mediums Act.

*A CELTIC CROSS* – *Pagan symbol of the Elements (opposite) – on the remains of an ancient oratory. Above, though the Dingle Peninsula was governed by the great deity Duibhne, relics of the Christian god now dominate the landscape.*

We have no clear record of the number of Witches of both sexes who were burned or hanged, but some estimates are as high as six million in Europe. At the base of Edinburgh Castle a plaque commemorating the Witches who were burned in its courtyard refers to 2000. The list of associated crimes was so huge that these executions were tantamount to active genocide. Those who perished included women in power, beautiful women, ugly women, single women, women who had lovers outside marriage, women who loved women, men who loved men, herbalists, midwives, women with property, women who were disabled and any who were not active Christians. Time and again throughout history there are examples of the Christian church imposing itself on the existing old religion, maligning its practices and distorting its vocabulary.

*A HIDDEN STONE* near a forgotten well dedicated to the Goddess Bride, once the most important deity in Eire. Although essentially a Fire Goddess, she was powerful enough to govern Water sites as well. This stone bears a small carved cross.

The word Pagan, for example, originally meant 'rustic'. Country dwellers were the last to be educated and the last to convert to Christianity. Heathen literally means 'men of the heaths' and again refers to those it was hardest to convert. Witch derives from the Anglo-Saxon *wicce*, a male Witch, one who shapes, gives shape to or is wise. Given that distinct cultures have always had their own religious practices, there was no reason why these people should have been interested in a spiritual system devised by another culture, nor, unless they were actively harming people, any reason why they should have been prevented from worshipping in their own way. The same is true of many aboriginal peoples today. It has been a hard lesson, a lesson learned through the annihilation of another culture, and yet we have seen it with the Australian Aborigines, the American Indians, African tribes, the Inuit peoples and the indigenous peoples of the Amazon region. The process carries on under many names, not the least of which is progress.

With the loss of these cultures we lose many medicines, many animals and plants, many wise ways and spiritual practices that enhance the planet. These forgotten cultures haunt us with their myths and stories. Once they are almost extinct we seem suddenly to get the point and realise what we have lost just before it is too late. This is the case with many of the early peoples of the British Isles. Their teachings were subsumed into the culture as epic stories. The Goddess is changed into a Queen and the tales are told as if they come from a distant time. We have lost the wise ways of the Celts in all but the stories. At the moment there is a resurgence in interest in the ancient teachings but the body of lore in Wicca and other Pagan sources is sparse. Much of Wicca has been reinvented in the 20th century with an eye to the ancient ways. Pagan historians have made a concerted effort to set some of the records straight and to include information from local customs to feed the greater picture of Paganism in Britain.

It was to the healing arts that early Witches were most connected. Spells that healed and spells that shifted consciousness are reported time and again in

Witchcraft trials. People went to Witches with psychological problems as well as physical ailments. Midwives were once hanged as Witches and so were herbalists. Yet the infamous Witch's flying ointment was made of a selection of herbs that are now part of modern medicine. Digitalis – now commonly used for heart complaints – is extracted from foxgloves. The wise women – and men – of the village studied the herbs that grew locally. They were aware of the existence of natural medicines and made use of poultices, tinctures, tonics and draughts. They produced simple, effective remedies from observing nature.

In Britain as in many countries, midwives are an integral part of the medical community and this move is accompanied by a greater respect for women's wisdom in the area of birth. Home births and water births are once again on the increase. For hundreds of years there was an overriding distrust of women being in charge of their bodies, but now we seem to be looking back in an attempt to create a greater balance and learn the lessons of the early peoples.

Medicine is only one example of the revaluation of ethics and treatment of disease. The modern rise in the interest in Paganism is no coincidence. It comes at a time when there is a widespread reassessment of the values espoused by the West as 'progress'. Some of the information held in oral traditions is now being taught and a web of Wiccans and Pagans offer this information in workshops and literature. Yet the word still needs to be explained and represented. We have for so long been fed images of hideous crones who exact some form of punishment on the innocent, or of seductive beautiful Witches who render men impotent, that the concept of the Witch as priestess is yet to be widely understood. The Witch as advisor, healer and wisdom-keeper is a path that many men and women feel able to explore, now that they are no longer actively persecuted. Yet by no means all such people are comfortable being 'out' in professional life, nor indeed do they tell the neighbours. We have come a long way since the burning times, but there is still a need for greater understanding and tolerance.

# STONE CIRCLES, RITUAL SPACE

Stone circles are carefully constructed ritual sites with specific purposes and entrances. In an attempt to discover them, whenever I visit a circle I allow myself to be guided by the land. I send questions into the atmosphere such as, 'Where are the natural pathways through the circle?' or 'Are there any sheep paths that pass through it?' In times when it was dangerous for people to be open about the old ways, only animals visited the more remote sites. Sheep, wild goats, wild horses and cattle all know the land instinctively and understand the subtle energies of the ley of the land. The thousands of years of animal husbandry that predate your visit to the stones will leave clues to the site's secrets.

The path through a circle is not always through the centre. Sometimes the centre has shifted and sometimes one stone 'feels' more central than another. If you visit a popular site there will undoubtedly be local experts who have their own ideas about which route to take to enter the stones and it is worth asking around. I had an experience of this at Lough Gur in Eire, at the largest of the three circles there.

I had arrived early one morning to have some time without the constant flow of tourists. The day was hot and windy and I looked for an indication where to enter the circle.

Because of the number of visitors who go there, this circle, like many others, has a pathway to enable local farmers to keep people from wandering across their land. So we are guided into the circles in a new way – by the invisible hand of the tourist authority, rather than along the ceremonial path that would have been honoured by those who created these mythic monuments.

I walked to the centre of the circle and gazed at the extraordinary trees that surround the stones. I was looking for a camera angle that would enable me to capture the magic of the place. The light was becoming too bright for an interesting shot, so I retired to what I thought was a natural break in the stones on the far side of the circle. It seemed the perfect place to photograph the stones directly opposite. I spent some minutes shooting different angles, but none was as good as the view I had from that little break in the circle.

*LOUGH GUR – the processional entrance to the stone circle (previous page). As the sun's rays shine on Midsummer morning, the light illuminates both the passage and the V stone (opposite) on the other side of the circle.*

While I was there, crouching in the stones, a man came towards me and introduced himself as the farmer who owned the land. 'You've found the processional entrance, then,' he said. The break in the stones, as it happened, was the old entrance into the circle, and the stone opposite this pathway, known as the V stone, captured the light on the morning of the Summer Solstice. The rays of the sun shone through the little passage and on to the V stone, which was aligned to the path the rays would take. In celebration of the solar deity, often described as male, the sun penetrated the circle, often described as the body of the female Earth Goddess. The ritual, usually interpreted as a fertility rite, concluded with the 'charging' of the stone whose position ensured the meeting of sun and Earth.

Charging is one of the most important aspects of magic. It refers particularly to the energising of an object by the sun's rays, the moon's light or human will, but it can also be used in connection with charging crystals with the energy of the ocean or the energy of fire. The concept of charging suggests that the object concerned has been transformed in the process, that after being

charged it has become different in some way. 'Charged' stones are believed to retain information in the form of an exchange of energy between the stones and the people who visit them in the course of countless generations. In the case of the V stone at Lough Gur, the symbolic act of the sun lighting up a stone that had been carefully positioned so that this would happen at Midsummer must have signified something to the local people.But herein lies the problem of interpretation. What we read into these ancient rituals is informed by our historical context and we can only draw what seems to us the most appropriate conclusion, leaving the answer open-ended.

We do know that there is a relationship between stone circles – made of carefully selected stones, laboriously moved into position, in some cases from hundreds of miles away – and areas named with feminine references. These great temples were made of the Earth but looked to the sky, as they were both sun- and moon-aligned. The positioning of the stones was carefully considered to enable the ancestors to calculate astronomical events. In Callanish on the Isle of Lewis, for example, there are times when the moon appears to sit within one of the smaller stone circles. At other sites, when the sun is at its height on the longest or shortest day of the year, the rays may enter the circle either directly or by casting a shadow into a circle, as is the case at Stonehenge. Archaeologists have interpreted circles and passage tombs which receive the light of the sun on specific days as being penetrated by the action of the sun moving into position. The interaction of the solar event and the positioning of the Earth temples may have been seen as an exchange of energy without being specifically male and female. There is a long tradition of the sun or sky being associated with a male deity and the Earth being female. If these traditions hark back to the previous Earth religions, then it is likely that these temples had some aspect of a fertility cult working within them.

The exchange of energy is visible when there is strong sun, but if the skies are overcast then our understanding of what is happening may change accordingly. These events may have been seen as omens of things to come; the strength of the sun's light may have indicated whether this would be a good year for the community, either in general or in some specific respect such as a bountiful harvest. The V stone at Lough Gur stood waiting to receive the sun, a constant in the ever-changing weather patterns of the seasons. It would have been key to the ritual activities taking place in the circle and the repeated act of 'charging' would keep the stone active in terms of the energy it gave out.

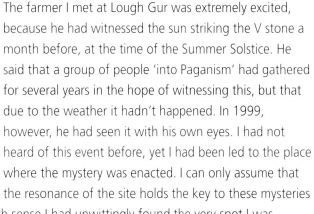

**ONE OF THE SMALLER**, *neglected stone circles at Lough Gur. Although tourists still flock to the main circle, this once important ritual site is left now for the ever-encroaching foliage and the few visitors who hop over the fence.*

The farmer I met at Lough Gur was extremely excited, because he had witnessed the sun striking the V stone a month before, at the time of the Summer Solstice. He said that a group of people 'into Paganism' had gathered for several years in the hope of witnessing this, but that due to the weather it hadn't happened. In 1999, however, he had seen it with his own eyes. I had not heard of this event before, yet I had been led to the place where the mystery was enacted. I can only assume that the resonance of the site holds the key to these mysteries and that with my sixth sense I had unwittingly found the very spot I was looking for.

If it is not possible to ask for information about entrances, you may be able to learn about stone circles using the old methods for unlocking the secrets of the stones. Plan your approach by looking from the outside and feeling the ley of the land. Look for symmetry and the size and shape of certain stones in relation to others. Entrances and exits, openings and closings, are key to the art of ritual. Stone circles, more than any other ritual site, become an outward manifestation of the ritual space. Ritual activity takes place in every culture throughout the world and predates any buildings; indeed, some of the earliest buildings became centres for the rituals. Most rituals take place in a circle – the shape of the space lends itself to ceremony and communication. The modern concept of 'casting' the circle is an indication that this geometric shape somehow offers a particular potency to a rite.

The parallels to the circle used in Wicca are obvious. Whenever I do a ritual it takes place in a circle which is drawn, or cast, by one member of the group, while the others lend their psychic energy to the act of drawing. This creates an invisible boundary between the people and events that take place 'in the circle' and normal life outside it. Once the circle is cast, whatever takes place in the ritual space is said to be 'between the worlds'. The creation of safe space is essential to the acts of transformation that are the magic. What takes place between the worlds can never be discussed, nor can it be a place of repercussion. What it is is a place of individual responsibility. Within the circle, individuals' true selves emerge and barriers are dropped. It takes practice to ease out of the outward persona – we have to learn to let go. But once we have achieved this, anything is possible in the sacred space.

There are many things to take into consideration before and after a ritual. Witches come 'properly prepared' to a rite. Cleansing and meditation are two favoured forms of preliminary work. Coming to a circle with the cares and worries of your day at work inhibits the ability to let go. It is essential to clear your mind and body of anything that will distract you. Preparation of the self and the space in which you are working can make an incredible difference to the potency of the spiritual work and it is messy practice to cut corners and forget. The same applies to the spiritual journey: before entering a stone circle, prepare yourself for the experience. Remain conscious, greet the circle as you approach it and ask how to enter. Receive the information in silence.

*THE BEAUTIFUL CROSS SLAB OF RIASC, Dingle Peninsula, stands on the site of an early Pagano-Christian community. The spiral motifs, less sophisticated than those at Newgrange, are a recurring theme in pre-Celtic Neolithic art*

Another way of entering an unknown sacred landscape is to welcome it into your heart and introduce yourself. This helps you to attune yourself to a place already steeped in ritual. Many have stood where you stand and worked the magic of the place. Over time, the energy of a sacred site acquires strength by repeated acts of spiritual practice and worship. If you allow yourself to listen to the land, to speak to the place in your own way, then enter the circle after observing and listening to your instinct, you will easily feel your way in.

Just as the mystery of any ritual circle takes place at the centre, so the centre of the stone circle is a key to the energy that resides there. It is the place of transformation which focuses your attention on the core of your being. Take time to hear the messages from the centre. Allow the flow of thoughts to move through your being and witness them without judgement. If possible, sit in the centre and meditate. Inner and outer centres become as one.

'CENTRING YOURSELF' is a term used in many spiritual practices to describe the process of calming that is the preparation to any magical act. Anyone who has done any kind of yoga practice, Qi Gong, Tai Chi, martial arts or breath technique will recognise the amplification of energy that occurs in stone circles. Once you have found the centre of the circle, centre yourself by breathing deeply into your belly. The messages from the centre of circles come from deep within the land and the body of the Earth Goddess. Meditations become powerful and healing is heightened, whether it be hands on or distant. Playing a magical instrument at the centre of a circle enhances the energy of the music. Perhaps the sacred geometry of the circular shape confines the energy raised within the stones to the centre, which acts as a receptacle.

# Stonehenge

The most famous stone circle in Britain is the great Pagan temple of Stonehenge. Legend tells us that Merlin, arch Druid and advisor to King Arthur, brought the stones from Eire by means of enchantment. They are in fact from Wales, brought over sixty miles by means of glaciers which carried the stones to Salisbury Plain. They were erected by means of roller logs, ropes and a great deal of ingenuity.

Stonehenge is believed to be an ancient observatory. The positioning of the stones in observatory temples provided the ancestors with a very accurate astronomical calendar. These may have been used to calculate the times of equinoxes and solstices which, being based on the solar calendar, occurred at precisely the same time each year. Later Celtic festivals such as Beltane and Samhain were 'movable feasts', depending on the seasons and possibly related to such variables as the first frosts of the autumn.

*STONEHENGE is an awe-inspiring ritual site which holds an enigmatic central mystery. It has the largest stones of any stone circle in the world, erected by people to whom the construction of such a great temple must have been of huge importance.*

The stones are part of a huge sacred landscape on Salisbury Plain. Nearby is one of the largest Neolithic barrow cemeteries in Europe. Cursi – long man-made banks and ditches – run along the land and enigmatic post-holes are dug into the earth. Within the circle is a grouping of huge megaliths that form a crescent womb.

The rites at Midsummer, when the play of light and dark in the stones becomes visible, take place in the womb-heart of the stones. The so-called King Stone stands outside the circle at an odd angle. When the sun shines at Midsummer the shadow cast by this stone moves out towards a space in the stones and enters the womb centre. Here light and dark exchange glances. Whereas at Newgrange passage tomb (see page 153) the ray of light penetrates the womb heart, here at Stonehenge it is the ray of shadow. The creeping shadows strike into the heart of the temple where rites of the sun were performed. The shadow cast by the huge King Stone sends its dark presence creeping forward toward the circle. When it enters the stones it can be read as a penetration of the veil. The solitude within is pierced by an outside presence.

We sense this event in tune with the stars and with the ancestors. Something touches us between the worlds and the cracks in our soul open to let in the light. On the day of most light, the shadow shifting between the worlds comes to rest in the womb of the Earth Goddess.

For me the circle is one of the formative sacred sites of my journey. There is

something so essentially Pagan here. It is a temple in which the sky meets the earth and the sun meets the moon. Although it is no longer possible to wander freely among the stones, you can book a special viewing at either sunrise or sunset on any day except the Summer Solstice, when the modern order of Druids is allowed to celebrate its rites. As a Witch I book my visits as private time in the stones with no obvious connection to my spiritual path. Here, in the shadow of the great stones, I celebrate the hidden mystery of the Goddess. On one occasion I stood with a circle of women inside the central horseshoe and we spoke the names of ancient Goddesses. Though it was a simple ceremony it had a potency that rose out of the land and the circumstances that have forced women to celebrate their path in private. The timelessness of the ancient temple sheltered us from the elements and there was a great sense of peace.

Stonehenge is also a national monument and a kept temple. When a temple becomes a tourist haven the sacred is partially lost, but at least the stones are preserved from decay. As it is not permitted to touch the stones and you are always observed by guards, the most profound thing I have done at Stonehenge is to lie on the ground and allow the image of the changing clouds and sky to form a background to these towering monuments.

*WITH MY FEET* on the earth and my eyes raised skyward, I feel a deep connection to the early Pagan celebrations that were held at Stonehenge. There is a pleasure in simply taking your shoes off and speaking the names of the old goddesses.

# The Nine Maidens, Boskednan, Cornwall

The circle known as the Nine Maidens is one of many that speak of a Goddess connection. The number nine recurs when it comes to stone circles. It is sacred in Wiccan lore for it is three times three. Nine is the number of months that it takes to bring a child into the world, representing three times three menstrual cycles within the womb. The three aspects of the Triple Form Goddess – Maid, Mother and Crone (see page 65) – are symbols that enable us to understand the passage through the distinct stages of consciousness in an individual's lifetime.

There are many reasons to believe that rites in stone circles involved great moments of transition in human consciousness. Birth and death were two key themes and the Goddess of the land, as seen in the womb or receptacle at the centre of the stone circle, was at the centre of spiritual expression. The active interest in birth and death as part of being human meant that fertility rites were an important aspect of the worship of the divine and were practised in all stone circles. The Great Mother was honoured with fertility practices. Lovemaking in stone circles was common, and people still make love in the stones as a resonance of the old rites.

As the image of birth, the child emerging from the womb was a powerful representation of woman as creator. The rites of passage of humankind necessitated an honouring of the fact that everyone came out of a woman. There

*BOSCAWEN, CORNWALL* – *women gather, as the ancestors did, at the central stone. This circle is used regularly by Cornish Witches and the stones are still 'active'. Witches were once numerous in Cornwall – there are many sacred sites within a few miles.*

must have been a sense of awe for the Mother Goddess, the protectress and healer. She who bore life out of her own body endured bloody rites that were seen, accepted and revered by everyone. Just as the rite of passage of death was seen, so too were births that brought death to the mother. This cycle of woman as birther, and the strange twist of fate that can cause death to the one who is giving forth life, was seen an aspect of the great Goddess. She gave with one hand and took with the other. Thus all that was momentous in human experience was encompassed in the concept of the divine.

## CORD RITE AT THE NINE MAIDENS

*The number nine, which is a potent number for anyone wanting to access the mysteries of the Earth Goddess, may work particularly well at the Nine Maidens, but this cord rite can be used at any stone circle, or in any other ritual space.*

*The tying of knots is one of the oldest forms of Wise Craft and is part of the Celtic tradition of rag magic. It involves the ritual knotting of a cord, string or thread and tying intent into the thread. The thread may be any colour that represents what you wish to address. Many colours signify certain kinds of intent – for example, red is used to speak of passion, birth or healing with fire. There are many symbolic systems of colour association that work with different intents. In Wicca, each of the four elements is associated with a colour, so it is usual to use blue or yellow for Air, red or orange for Fire, green or blue-grey for Water and black or brown for Earth.*

*If you follow a different elemental system, use the colours that you have learned or follow your intuition. For some people green means anything to do with the Earth or environmental concerns. For others it symbolises money. The intent is what is important and the knots are a way of tying a wish into the string.*

*The layout of stone circles enables the ritual to be led easily by the site itself. To stand before a stone and state your intent, and then to move to the next to make another statement is a powerful way to do knot magic.*

*Standing at the centre of the circle, light some incense. As the smoke drifts up, begin to breathe in white, clear light and energy. As you breathe out, breathe out*

*THE NUMBER NINE* also leads to many stories of Witches. They relate to simple rituals that allowed *'the individual to become a witch, to gain access to ghosts; the devil; or the Underworld'.* [1] There is a plethora of Pagan imagery with Christian description. Many groupings of nine, especially groups of women, appear in historical or mythical contexts as a unit having supernatural powers.

In the 19th century a man named Hunt wrote about the presence of Witches at Zennor, a village in Cornwall not far from the Nine Maidens. At Midsummer they met on the farm-covered hillsides and *'lit fires on every cromlech, and in every rock basin, until the hills were alive with flame'.* [2] Amidst these rocks, there was one called the Witches' Rock, which was where the Witches gathered at midnight.

After the rock was removed, Hunt noted that all the Witches had died and the faeries fled. He regretted the disappearance of the rock because *'anyone touching the stone nine times at midnight was insured against bad luck'.* [3] It was also the local belief that any woman could become a Witch by touching the stone nine times at midnight.

all the blocks and thoughts that are muddling your clarity. This takes as long as it takes. Keep the intent firm as your breath flows in and out. Allow the image of your body and mind to free itself of all unwanted thoughts and then let the blocks dissolve with the exhalation of black smoke. Allow the inward breath to match the outward breath. Do not force it. The balance comes out of gentle yet focused breathing. The thoughts that drift in will drift out with the black smoke and the light will fill up your body.

It is here in the centre that you state your overall intent. Out of the silence of breathing, focus on what you are tying into the thread. Say it out loud or silently in your mind. It may be that you wish to bless a child who has come into the world and you have come to the circle to make a blessing necklace. It may be that there are nine wishes you wish to give to someone. Equally, there may be nine obstacles that you wish to remove. To release an issue or a bad habit, try repeating the same intent nine times or speak of nine different intentions on the same theme.

When working to let go of something – if you are trying to give up cigarettes, for example – it is often valuable to remove any obstacles first and then to work for nine affirmations that bring healing. To remove the obstacle is to identify the underlying problem. Then it is important to fill the void with positive energy and willpower. Once this is done, the cord may be buried, thereby burying the obstacles of which you wish to let go. When you feel that the clarity has returned to your body and mind, choose the stone that receives you. Ask which direction to go and go to that stone.

This is an easy rite to do whether or not there are other people at the site. Sometimes privacy is not possible and unless you are visiting at night other visitors are unavoidable. Silence is one of the most powerful ways to work and is the key to the element of Earth, whose sacred doctrine is 'to

**FOR THE BIRTH OF A NEW BABY** *it is nice to give parents a blessing candle. You can start the candle off by carving your intent into it, and then the parents can light it to activate the blessing for their child. When carving sigils (signatures) into the candle be in a conscious state with intent and seal the blessing with water or wine.*

*It is also common to charge a candle by passing it through the four elements. If you are able to charge a candle at a stone circle, make sure you know which direction is which, take the candle towards each and carve something for the qualities of each (see page 39). Then use the centre as it has been used for centuries by declaring your intent there.*

*As well as cords and candles, it is possible to charge a crystal with energy. This is also true for a rock, a bulb, a gift, a clootie (rag), a wand or any Witch's tool.*

*keep silent'. Whether you are speaking the spell aloud or saying it mentally, with the thread in your hand make your way around the circle and knot it nine times.*

*Face the stone and hold the cord in your fingers. Speak directly to the stone and state your intent. For a child whom you want to endow with a number of blessings, use the nine stones for nine specific wishes. This may be as simple as 'I bless this child with strength' or 'I bless this child with health.' For a woman who is about to give birth, bless each knot with the same intent, that of easy birth. It is this part of the practice that sets the intent in stone. Speak each statement of intent directly, without wasting words. As the words pass your lips, knot the thread firmly.*

*Knot the middle of the cord first, then the left end, then the right. The rest of the knots fill the spaces in between. Once you have nine knots turn and face the centre of the circle and go directly to it.*

*To consecrate the cord speak your intent at the centre and spill some water or wine on the cord. If you have visited a sacred well you may wish to bring a bottle of that water. Many sacred sites are not far from one another and the pilgrimage to the well can also be done with intent. If this is not possible use the old 'hag's seal' and simply spit on the cord. This done, the cord may be tied to a wrist or given as a present to wear as a necklace.*

*The cutting of the cord is an important aspect of the rite. To end a spell you may choose to do this. Once cut, let the cord do its work. Don't focus on it. This is an important aspect of any magical working, as important as the magic itself. To hold a thought is to keep it chained. Witches and shamans let the magic go on its course without interference. To bring doubt into the equation is to undo the valuable work you have done.*

*THE ELEMENTS* *are one of the foundations of any magical work. They symbolise some of the earliest beliefs of what life itself consists of. In ritualised representations of sacred space each element is associated with a direction; the space in the centre of the circle is the fifth and most important. This is Aether or Spirit. André Breton, the leader of the Surrealists, called it* frisson. *The sum of two ideas is greater than the individual ideas put together. The centre of the circle is the place of the idea, the possibility, the exchange, the unnamed, that which cannot be spoken, the heart centre and the magic. It is the sacred space between the worlds where anything can happen.*

*This theme is repeated time and time again, throughout many cultures: American Indian, Wicca, Pagan circles, Thelema, Western Paths, the Celtic map of Eire. It is not the only way to conduct a ritual, but it offers a diagram within which to work. The circle is the protection and the site in which the magic will be worked. Each of the four directions has a guardian or qualities associated with it, offering the possibility to look outward beyond the circle's bounds.*

*In addition to the idea of a circle that is flat and then goes off in the four directions there is also the perspective of the underworld and the overworld. The divine in the West often exists above, but has been historically perceived both above and below. Nowadays we associate above with the heavens and below with the dark realms of the underworld. The underworld has become a place of shadows but was not always so.*

*When doing rites of protection it is important to cover all angles, including above and below. There is much in psychology to suggest that when heaven became good and hell bad, there was an active attempt to conceal the shadow world and face only the light. In Wicca this perspective is challenged to include both dark and light, up and down, like the yin/yang of Chinese philosophy. To ignore the hidden depths of the*

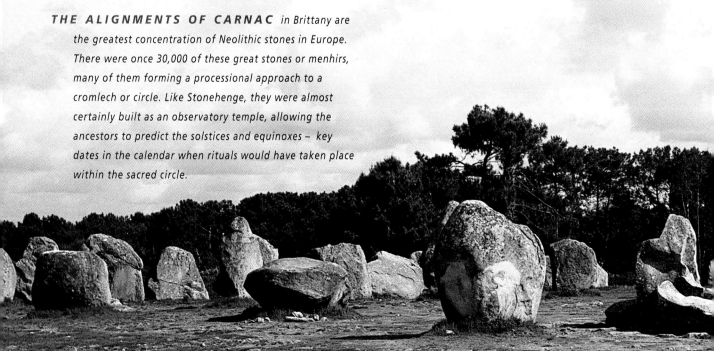

*THE ALIGNMENTS OF CARNAC* *in Brittany are the greatest concentration of Neolithic stones in Europe. There were once 30,000 of these great stones or menhirs, many of them forming a processional approach to a cromlech or circle. Like Stonehenge, they were almost certainly built as an observatory temple, allowing the ancestors to predict the solstices and equinoxes – key dates in the calendar when rituals would have taken place within the sacred circle.*

*personality leaves one open to the destructive tendency of looking only for what is obvious, light and good. Actively to face those parts of the personality that are not so light is to bring a balance that is an essential part of being human.*

*By working with the qualities connected to each element it is possible to address areas of life and self that are not in harmony. Each element also has its own doctrine. The doctrine of Air is To Know; Fire, To Will; Water, To Dare; Earth, To Keep Silent. The combination of the four leads to something greater than the four individually. Aether is the centre and it is To Go.*

*To give some idea of how the elements are used it is useful to think of them as a balancing tool. If any crystal, cord or magical talisman needs to be activated or charged, it can be passed through each element to bring a blessing. Once the object is passed through each it is considered consecrated and this is a way of activating its intent.*

## AIR

*all issues of communication, literature, photography*

*the bigger perspective*

*breath and breathing*

*clarity, intelligence*

*flight, change*

*laughter, music*

*community*

*Sky Father*

**symbols of air**
*incense, the athame or Witch's knife*

## FIRE

*all issues of inspiration, passion, leadership*

*warmth, healing, lust*

*male energy, protectiveness*

*direct, overt activities*

*study, letters*

*courage*

*issues of the heart*

**symbols of fire**
*fire, a candle or wand*

## WATER

*all issues of emotion, love, spirituality*

*female energy, womb, menstruation*

*nurturing*

*birth, healing deep wounds*

*gentle flowing feeling*

*pleasure, happiness*

*the blood of life*

*issues of the soul*

**symbols of water**
*water, cauldron, chalice or cup*

## EARTH

*all issues of the material realm*

*the home*

*family, loved ones, friends*

*solid foundations or foundational work*

*work, career*

*Earth Mother, healing through nurturing*

*feasting, the harvest*

*abundance, prosperity*

**symbols of earth**
*soil, rocks, crystals, the pentacle*

## A CHILD'S NAMING CEREMONY

*A few years ago, I had the opportunity of using the knotted cord at a young friend's naming ceremony. I had known Shamal since she was born. When she was about to turn seven, her mother, who is my good friend, suggested that I do a naming ceremony for her. We thought that she was ready to mark a threshold. The practice among Native American Indians is that the child chooses a name or has a name chosen for them, so we asked Shamal to choose a special name for the rite. Wiccans do a child's naming ceremony earlier.*

*The group that assembled was quite large and they stood round in a circle. The circle had been cast and the quarters invoked. The quarters are represented by the four directions and also symbolise the elements: Air, Fire, Water and Earth. In Wicca, Air is associated with the east, Fire with the south, Water with the west and Earth with the north.*

*Before performing a rite, Witches attempt to use the structure of the circle and quarters to make a safe place for transformation to take place. An invocation of each element at the quarters is spoken to bring awareness of the importance of the balance of the qualities of Air, Fire, Water and Earth, so that at the centre of the circle all creative energies will be attuned.*

*At one of the key points in the rite I passed round a cord. Shamal took the cord to each person in the group and faced them. They each tied a knot and wished something for her, and she was left with a wishing cord. Once everyone had knotted it, Shamal passed it through all the elements. She held it over the incense, which represents Air, then passed it through a flame, sprinkled water on it and finished by putting some earth on it. At the centre we did magic for the ease of passage that each stage of life brings. We charged the cord with love for the journey ahead for this young girl.*

*Recently Shamal attended another ceremony. This time it was marking a coming of the blood rite for some of her girlfriends who had started menstruating. For this rite they passed a wishing cord around and then a Crone woman sat at the centre of the circle and told them stories of womanhood. The Crone passing wisdom to the Maid is one of the important relationships of the Triple Form Goddess. She lends a word of wisdom to the younger women on their path.*

*MEN-AN-TOL HOLED STONE* *The hole is seen as the threshold of life and death or the birth canal from which all life emerges. To this day women crawl through it in the hope of becoming pregnant. Fertility and sexuality are the essence of this holed stone.*

# Men-an-Tol Holed Stone, Cornwall

Men-an-Tol is another site of standing stones, one of which is round with a hole through it. Holed stones of varying sizes are seen throughout Britain, but this is a particularly beautiful one.

Not surprisingly, Men-an-Tol has long been known as a site of fertility and healing. Women who want to become pregnant still crawl through the hole, and various other healing powers have been attributed to the stone:

*The one at Madron [Men-an-Tol] is sometimes called the Crick Stone. It gets this name because in days not very long ago people afflicted with rheumatism, sciatica, etc, in May, and at certain other seasons of the year, crawled on all fours nine times around these Men-An-Tol from east to west, and, if thin enough, squeezed themselves through the aperture. This was then thought such a sovereign remedy for these diseases that parents brought their weak-backed children and carried them around. To work the charm properly there must always be two people, one of each sex, who stand one on each side of the stone. The child, if a male, must first be passed from the woman to the man; if a girl, from the man to the woman, and always from the left of the one to the right of the other.*[4]

Men-an-Tol is also cited as a place to work a tuberculosis cure. Children were passed naked through the hole three times and then drawn on the grass three times 'against the sun'. The inherent belief that the passing through a hole brings about new life is the link to the ancient Goddess as life itself. Babies are still passed through this beautiful stone as a blessing to bring them good health for life.

**DOLMEN ZÜSCHEN, KASSEL, GERMANY** *The megalithic monument of Züschen is a collective gallery grave excavated in 1898. Some 40 human skeletons have been found here. The stone slabs on both sides of the grave are covered with rock carvings, and the holed stone must have been a symbol of fertility, perhaps showing a link between the various stages of life which culminate in death.*

*THE HOLE is woman's hole of wisdom, the gateway to the Earth Goddess. The same imagery is found in sheela-na-gigs (see page 112), the female gargoyles carved into doorways and pillars of churches throughout Britain. The doorway is symbolic of the threshold over which we pass in order to gain wisdom and it is analogous to the birth canal through which we all must travel to manifest on this Earth. These little figures hold their pussies open in an outrageously provocative manner. Almost certainly linked to Goddess worship, sheela-na-gigs are early representations of a sexual woman. They are meant to be frightening and to keep evil out of the church. The masons who built the churches were clearly Pagan and carved these images while employed by the Church. It is part of the hidden history of Paganism that we find Green Men, phallic imagery and symbols of Goddesses in the rafters of churches. Many sheela-na-gigs have been destroyed because of their sexual nature and because they hark back to Goddess veneration and birth magic. They are part of the female-focused mythology into which holed stones fit. Primitive and obvious, the hole and the birth canal are as one in the image of the holed stone.*

In Cornwall the holed stones, which are known elsewhere as hag stones, were called adders' beads and as late as the 19th century they were used as magical amulets against 'evil Witchcraft and mischief by the small people'. This is another example of Christian language being turned against people of the old religions, for in ancient times the amulets would have been given as protection *by* Witches or cunning people – possibly to guard against Christian persecution.

Small holed stones have long been central to rites of passage. Milk and blood were poured into the hollows of stones. Engagements were sealed between lovers. On the Dingle peninsula in Eire there is a stone that acted as the site for agreement of a contract. Anyone holding another's hand through the stone made a firm commitment and the breaking of the oath would result in the offender being outcast. There were many such sacred ceremonies of sealing the pact: stones are used time and again to make the issue solid. Arthur pulled the sword Excalibur from the stone to seal his kingship. The stone that for centuries sat under the throne of England, the coronation stone or Stone of Scone, was recently sent back to Scotland, from where the English had stolen it. Power being granted by means of a stone is an ancient symbol and occurs throughout Celtic mythology.

## ELEMENTAL PATHWORKING

*This pathworking is a method of balancing the elements in your life. There are many ways to do this, but this one works on a symbolic level, with references as old as the Celtic lands. The symbols of the mermaid and the eagle make it especially appropriate to the sacred sites of Orkney, though it can be done anywhere.*

*The response you have to each stage of the journey may give you some idea of the messages that come from each element and indeed your subconscious. It is a shamanic tool used by Witches to attune themselves to the universal pulse.*

*It is possible to use this pathworking in a group with different people reading different parts to enable the others to take the journey. If you are on your own, read it with a lazy attention. Allow yourself to go into the story and do what is required. There will be parts that you can just hear and parts to remember. I use drums, rattles and Tibetan bowls to take people into pathworkings, but on your own you can do it by simply drifting through the imagery.*

*Start the process by making yourself comfortable, either sitting or lying down and breathe very deeply. Anyone who is acquainted with yoga and the deep belly breath may wish to use this technique, as it creates instant relaxation. If you have not learned yogic breathwork then simply breathe through your nose. The breath is relaxed and deep. It moves into your belly so that if you put your hand on your stomach you can actually feel the rise and fall of the breath. The sound of the breath is a slight hissing or snoring in the back of the throat, for though it is through the nose, the focus is in the throat. Once you have become comfortable with this breath, allow the in breath to be in equal measure to the out breath. Nothing is forced and the rise and fall is continuous. There is no hesitation between breaths.*

*Allow all the cares of the day to leave with the out breath and work your way down your body, releasing tension from each part as you focus on it. Release all tension from your jaw. Allow your tongue to hang heavy and unimpeded in your mouth.*

*Imagine yourself in a stone circle. It may be one that you know and wish to work with or a creation in your mind's eye. You find yourself in the centre of the circle.*

*You are lying on warm grass and the dawn is breaking. As the sun rises, the light filters through the stones and you hear the sound of birdsong. The warm morning sun creates a sparkle on the grass as beads of dew shine in the light. You are very relaxed and comfortable, as if waking from a deep sleep. You feel the tingle of a new day and the pale yellow and blue of the sky awakens you gently.*

*The stillness of the circle and this ancient land is pleasing to your newly awakened senses. You stretch and, as you do, a soft wind comes up and ruffles your hair. You start to wake more and breathe the fresh morning air deeply. There is a feeling of aliveness in this place. As you sit in the centre you hear the sound of a woman singing. There is a tinkling of bells and a strange mystical sound. As you turn toward the sound you see a flash of something passing between and behind the stones. You cannot see clearly who or what it is, but you hear a rustle of feathers. Suddenly a bird rises up from behind a stone and flies to the other side of the circle where it sits on another stone, watching you.*

*The bird is a huge majestic white eagle. As it watches you something happens and you start to see through the bird's eyes. It asks you to fly with it, to shapeshift and see the land from far above. You suddenly find that you have feathers and you are perched on a stone, looking at yourself in the centre of the circle. The bird tells you to fly and you take a leap into the air and find yourself flying easily above the stone circle. In the distance you can see the ocean and the green of the land all around. From here you look very small in the circle and in relation to the land. The bird tells you that it is giving you the gift of perspective. As you fly overhead all your problems are small and insignificant. The circle is ancient and in the centre you see a tiny person. This person is you.*

*The Eagle then asks if there is anything you have to tell that tiny person in the circle. Speak out loud. Remember. Presently the Eagle tells you that it is time to go back to the centre and you find yourself plummeting through the air. You swoop down above yourself and then find yourself in the centre of the circle once again.*

*The day has moved on and the sun is warmer now. The dew has gone from the grass and the deep clear blue of the noon sky is above you. It is hard to look up into the sky, for the sun's rays are too bright. You feel the warmth of the sun and the dryness in your mouth. It is getting hotter. As the heat penetrates your body you feel all your muscles relax even more. The sensation of sweat glistening on* your skin feels welcome. You are somewhat sleepy in the sun and you feel lazy and comfortable. The stones are giving off heat now and there are beautiful crystals shining in the light. The array of colours gleaming from the stones is the spectrum of the rainbow.

Suddenly you hear a beat of drums across the land. From somewhere in the distance you hear the heavy beat of hooves pounding towards you. Your heart races, but you feel sure that you are safe in the circle. From the south you see a white mare galloping towards you. She tosses her mane and comes to a standstill in front of you. As she looks at you her gentle healing eyes are captivating and she trots up and licks you on the face. The sensation of her tongue makes you laugh and she plays with you, rubbing her large head on your body.

The mare asks you to join her for a ride and invites you to get up on her back. As you jump on to her back she stands still to make sure you are ready and then races out of the circle. You ride across the landscape and she takes you to the hills in the distance. The land is soft to ride on and it is easy to stay on her back even though you have never ridden bareback before. The mare tells you that she has lived in the hills for centuries and is one of the wild horses who cannot be tamed. She has come to take you to a place where you can receive healing. She stops at a clearing on the side of a hill where there is a fire blazing. She gently lowers her big body so that you can slide off her back and tells you it is time to take all your clothes off and put them in the fire. She asks you to throw any sadness or stress and burn it in the blaze. Then she trots over to a small stream that comes off the mountain. It is time for you to drink of the blood-red stream that cascades down from the mountain. As you drink you are filled with a sense of well-being. The liquid, heated by the noonday sun, fills your body with a warm glow. The mare tells you that this is the blood of life that gives courage. She asks you what are your passions.

You tell her of the things you would like to do but have not done yet. Breathe and remember.

Now it is time to go. Naked and refreshed you leap up on the mare's back and she rides easily across the landscape. The heat of the sun keeps you warm and presently you return to the circle. The mare bids you goodbye and tells you that if you need her she may be found in the hills. Then she trots off into the distance.

*The day has moved on to the late afternoon. You relax in the circle and lie down on the warm grass. Clouds flit across the hazy sky. You stretch and enjoy the feeling of your skin in the air. As you breathe you hear bubbling water. It sounds like a song and seems to be coming from below. You listen to the lull of the water and it soothes your inner being. After a while the song becomes distinguishable. It is like an ancient ballad. It is a deep lilting voice and as you listen, the voice invites you to come to the waters. At first you cannot see where the water is, but as you walk around the circle you realise that it comes from beneath your feet.*

*As you track the brook it seems as if you are being led through the centre of the circle to the west and there you see some lilies growing near one of the stones. The lilies are a beautiful cream colour and the green of their leaves is the green of the faery lands. As you move closer to the stone with the lilies you realise that there are some ancient worn stone steps leading down into the ground. You hear a voice inviting you to walk down the steps. As you start the journey into the earth the bubbling and rushing of the water is growing louder. There are nine steps down and as you peer down into the darkness you see a beautiful natural spring. The spring runs into an underground stream that disappears into the dark earth, but is big enough for you to dip your feet in and wander for some way under the stone circle. As the stream meanders along you find yourself in a deep underground cave with a beautiful pool before you. Some light coming through cracks above the pool illuminates the cave. There in the pool is a woman half submerged in the water. As you watch she comes up on to a rock in the centre of the pool and you see that she has a fish tail.*

*The mermaid invites you to bathe in the pool. It is her magical voice that you heard before, when you were lying at the centre of the circle. She sings an enchanting tune that eases your body and spirit. As you bathe she combs her long flowing hair and tells you to look at your reflection in the pool. As you gaze into the pool you see your beautiful face, with all its lines and history. You watch as the lines disappear and the youth comes forth and then fades and ages. The mermaid sings of the three ages of woman and tells you that in each stage you find the other. Each line tells a story. She invites you to drink of the water and tells you that this is nourishment for your soul. She asks you what you most love about yourself.*

*You answer. Breathe and remember.*

*The mermaid tells you that the pool is a mixture of salt water and fresh. Here in the deep earth the spring merges with the ocean and she will swim out on the rising tide. She tells you to lie back in the water and she will swim back with you to the place you entered. As the water begins to rise up you let go and feel firm hands and fins take hold of your body. You flow along with her. The journey is easy and you feel a deep love for this helpful friend. As quickly as you were in the cave you are back at the steps and she bids you goodbye. With one great flick of her tail she disappears into the water and you make your way back up the steps. The nine steps feel longer and as you reach the land you see that it has become dark outside.*

*Back in the circle the moon shines down upon your naked body. The warmth of the air is still with you and as you go back to the centre you see a long cape made of soft wool. The cape, it seems, is for you and you wrap it around your naked body and feel warm and cosy inside. The moon is full and the hazy blue light filters down amongst the stones. It appears as if the stones are moving but when you turn to look they seem to stop. The stars twinkle above you and as you stare up into the sky the moon rises above the tallest stone in the circle.*

*You hear a rustling and a gruff snort. A huge white cow is heading slowly towards you, threading her way between two of the stones. Her gentle plodding pace moves in rhythm with your heart, which suddenly sounds very loud. You watch as she comes towards you and hear the heaviness of her step and the deep breathy sighs. The cow comes to stand by you and tells you to milk her. It is easy to milk the cow and you put the milk in a bowl that has been left for you at the centre of the circle. She invites you to drink and you put the bowl up to your lips and drink deeply of her milk. It is like nectar and feels good for you. She then asks you to remain silent and wait.*

*It may be something in the milk, but as you wait and watch, the stones begin to dance. As each moves with the next they form a circle. Writhing this way and that, it seems they dance to a heartbeat. Sometimes they look like women, sometimes like men, sometimes hermaphrodite, sometimes ragged, sometimes soft Bodies merge and transform. Sometimes you think you know a face but as quickly as you do, it is gone. The stones work their way faster and faster, becoming more and more animated. The ancient ones ask you to join them. You move your body easily between two of the nearest stones and take their hands. The stones pull you gently but firmly around and around. They become ecstatic and you laugh as you move. Faster and faster until you are being lifted off the ground.*

*You turn to the next stone and she is the Queen of Faerie. Her minstrels play wild tunes and you dance under the moonlight, feeling free. Your body feels so easy in the dance. As the night wears on you feel the desire that runs through your blood fuelled by your heartbeat. The dancers eventually stop and invite you to a feast. There in the centre of the circle is a magnificent feast of all the pleasures of the harvest. You eat and drink and make merry. The Faerie Queen tells you that she has a special draught that enables you to return to this place whenever you wish. She offers you a chalice of purple liquid and you drink it. The liquid is delicious and you think that you have never tasted anything so wonderful. There is a sensation and the liquid moves down through every pore of your body. It feels good and you smile at the Faerie Queen. She says it is time for her to go and kisses you. As she does, everything seems to dissolve and before you, the stones glisten with crystals. There is a light shining from each stone and as you watch, the faeries fade and the stones remain.*

*You are lying in the centre of the circle and the night has passed. There is a glow in the sky that heralds dawn. You feel very complete and peaceful. All the muscles in your body are relaxed, as if they have been worked in healthy exercise. As you watch the first rays of dawn pass through the stones and you feel yourself stirring. It is time to wake up now.*

*You gently stir your fingers and then your toes. Start to bring your attention back to where you are. Take a deep breath in and out three times. Really let the breath leave, and open your eyes.*

# THE EARTH GODDESS AND THE GODDESS OF THE LAND

The Goddess as Earth Mother is a Goddess who moves through the land. Long before there was agriculture there was the land. Five thousand years ago the people of the British Isles were as intelligent as we are now and had the same brain capacity. Unlike us, they looked to the land for inspiration and teaching.

The relationship between women and the land is as old as the hills. In the eyes of the ancients, the Earth Mother's springs sustained life. Birth was always woman's domain and women were perceived as the activators in the process of giving forth life. The springs were likened to pussies, the hills to her breasts and full pregnant belly. She was in the land. She was the living, breathing, cyclical Goddess. When she took her sun lover into her newly ploughed fields the crops grew.

We need only look to the land to feel the inspiration that the interplay of seasons, weather and landscape gave to the ancestors. The ancient Egyptians shared with the people of these isles the view that the land was alive and that ritualising the relationship and creating personified entities brought a more direct attempt to please, pacify, adore and honour the Gods and Goddesses. Just as the great Nile of Egypt was alive and it was necessary to perform rites to ensure the rise of her fertile waters each season, so the ancestors enacted seasonal festivals to encourage the health of the land and the people in Britain.

The remnants of these earlier belief systems are seen still in the local harvest festivals of rural Britain. The widespread practice of making a corn dolly from the last sheaf of wheat to be cut from the newly harvested field is a Goddess tradition. The last sheaf was especially significant in these ceremonies, as it was the last visual reminder of the whole harvest. This emblem of a good harvest was treated with special honour and was fed to animals, ploughed into a new field or made into a human figure and kept in the home. The corn dolly was a sacred object, representing the abundance of the land.

These folk customs are the descendants of earlier Earth rituals and speak in ritual language. The Earth as Goddess was honoured with temples made of stone which enabled the ancestors to enact the life-giving rituals of fertility and harvest in a specifically sacred site. Historically, men and women have always looked for guidance from the environment and used local examples of what constituted 'the masculine' and 'the feminine'. Most sites ended up being considered either male or female, but this did not discourage either gender from speaking to the site, nor did it prevent the site from 'speaking' to all.

**NOTCHES ON THE STONES OF CALLANISH** *(previous page and opposite) form viewing points for important astronomical events. This page, a Brid Dolly, also known as the Bride Doll. Bridgit's cross is the later Goddess in corn dolly form. In Eire these sacred objects are put in homes and considered lucky.*

## *Callanish as Earth Goddess, Isle of Lewis, Outer Hebrides*

*There are six rings of standing stones in the Callanish area. One is a true circle, two are elliptical and the others are deliberately built with a flattened side. Prehistoric rings in these shapes exist all over Britain.* [1]

The most lasting mythology in Callanish is that of the land as the Goddess incarnate. The Goddess of the land looks over these sacred and connected sites. Her form, the range of hills known as the Caillech or Sleeping Beauty, is a backdrop to one of the smaller circles. When you view the hills from a certain angle outside this circle her body seems to be contained within its perimeter. She lies in a gentle recline with her head on pillows. From one of the circles she may be seen to take the phallic stone in her mouth. This view may be seen as part of the interplay between the life-giving Goddess and the God whom she aroused with her breath. From her womb is born the moon; it rises across

her belly and drops down into the pillows above her head. From another viewpoint her breasts and belly suggest pregnancy.

The sacred landscape of Callanish was explained to me by Margaret and Ron Curtis, the local archaeologists. They know the secrets of the astronomical alignments and the Goddess who lies dormant in the land beyond the loch. Margaret has some fabulous photos of the moon sitting within one of the circles of Callanish. She showed me an image of the woman or priestess in the moon created by someone standing in the centre of the circle, arms spread. This happens every 18.6 years and will happen again in 2004 at the Summer Solstice.

***CALLANISH AT TWILIGHT*** *– the circle with the Maid, Mother and Crone stones. The colours of the stones correspond to modern Wiccan symbology of the Triple Form Goddess, and the white quartz of the phallic Consort stone glistens like semen.*

I met these two wonderful characters after a strange encounter at the main Callanish stone circle. Photographing the stones one morning I met a man who actually builds stone circles. He is a Tibetan Buddhist living in Seattle and, once he realised I was interested in the mystical aspects of

the site, he suggested I visit the keepers of the secrets of the stones. Though he uses modern machinery to lift the stones, he is building according to archaeo-astronomical plan. This is the study of ancient astronomy, but it is part of the old system of knowledge. In 3500BC Callanish was apparently famous as far away as Egypt, known as one of the finest observatory temples of that time.

Ron and Margaret had excavated a cairn with a cist (a small square chamber that holds the bones of several people) in the centre of it within sight of the stone circle. Despite their efforts, a road now runs through half the site. Nevertheless it is possible to stand on half the cairn and see the cross-section from the road. According to the Curtises, the cairn is connected to the main site and is the site of death. They believe that this was the passage a dead body would make across the sacred landscape – from the cairn along a carefully planned and possibly snake-like processional route to the great Callanish circle.

A medium had told Margaret that, whereas the cairn had been for the placement of interred parts, the circle was used for many ceremonies but was the place of 'sending' in ritualised form. Another circle, she said, was for the 'empty' remains. Once the remains of the dead had been ritually honoured, to let the soul go, the ashes and bones would be taken to a smaller circle and left.

Margaret and Ron have hypothesised that Didorus of Sicily, writing in 55BC, was referring to Callanish when he recorded the legend of the Hyperboreans. They were the people who lived on an island beyond the North Winds.

*There is also on the island...a notable temple which is...spherical in shape... The moon, as viewed from this island, appears to be but a little distance from the earth... The God enters the earth every nineteen years... The God...dances continually the night through from the vernal equinox until the rising of the Pleiades.*[2]

The reference to 19 years is close enough to 18.6 to be the moon seen in the small circle. This could mean that the moon was perceived as male and a god. If the symbolism of the Heavens interacting with the Earth was true of most of the sites in the area, then Callanish, like Avebury, was a landscape in which the individual could experience ritual thresholds of human existence at different sites.

The Goddess as a white cow is a feature of the mythology of Callanish. It is a symbol of nurturing and protection and has the gift of never-ending milk.

*During a winter of famine, following a Viking raid on Lewis, a woman met a beautiful white cow coming out of the sea. The cow spoke in Gaelic and told her to bring all her neighbours to the Callanish Stones, where each could take a pail of milk from the cow. Miraculously, the cow was able to give one pail of milk each day to each person, no matter how many came – and this eased the famine. The cow permitted one woman to fill two buckets, as one was for a sick friend, and this gave a witch an idea. She brought two pails for herself, but the cow would not allow the witch to milk her. So the witch came the next day with only one pail, but she tricked the cow by having a bottomless bucket, so that the milk ran to the ground. In this way, she milked the cow dry, and the cow was never again seen at the stones.* [3]

**THE CENTRAL CIST** in the large circle at Callanish. Cremated remains are known to have been buried here towards the end of the time when Callanish was 'active' as a sacred site, over 2000 years ago.

Often in tales such as this, the Witch is seen as the trickster, the 'bad guy'. Both cow and Witch are symbols of Goddess worship and both appear at sacred sites. The link between the prosperity of the cow and the Witch as thief came at a time when Christian beliefs were replacing the native Pagan religions and it is interesting that rather than the cow and the Witch working together they battle it out. In any case, the myth tells us of the departure of the practices of a previous Goddess-worshipping cult. The cow is literally milked dry and leaves the stone circle. In Hinduism the white cow is sacred and to this day is seen as a nurturing protectress. Milk rites, such as the offering of milk to the Goddess as a libation, are still practised. As well as cakes, colourful dust and honey, milk is poured over shrines in all parts of India.

In the Celtic myths the repeated reference to milk may suggest a link to ancient milk rites. The importance these stories place on mother's milk and the connection between the Earth Mother nurturing her children and the White Cow Goddess are indications of the esteem in which women were held in the Earth religions.

We are also told of the importance of Midsummer and Beltane (1 May) in the rites practised at the stones. Certain families who were 'of the stones' visited the circle on Midsummer Day to form engagements and it was on the Midsummer sunrise that 'the shining one' would walk up the avenue heralded by a cuckoo's call. This character may have been a priest or priestess who represented the sun; alternatively it may have been part of the faery tradition. Shining is a word that is often used to describe faeries or Pagans. The cuckoo was also supposed to call the Druidical May festival and gave its first call from the stones of Callanish. On May Day all fires were put out on the island and new ones lit from a central fire started by an old priest. It is also recorded that the Christian ministers expressly forbade anyone 'of the stones' from visiting on the old festival days. The Protestantism practised in the Hebrides was actively opposed to Paganism and the Church would have felt threatened by any locals in the habit of worshipping at the ancient sites.

I visited a Beltane site in the same area, Berna Bridge. According to Ron, if you line your eyes up in the direction the stones are pointing, the sun and moon will touch the hills on the horizon at certain times in the year and boulders have been placed on far mountains as markers. The Three Weird Sisters – as I call three stones at this site that twist towards due south – mark the most southerly point in the moon's journey, the time when the elliptical shape of its orbit brings it closest to the Earth. At this time it is possible to see the moon sitting on the mountain in that direction. The space between the stones marks the direction of the May Day sun, which is why this is considered a Beltane site. There is also a stone across the river, connected to this smaller site, which Margaret is sure is a male stone. Again, there is the tantric interplay of the stones from one site to another, symbolically representative of the interaction of male and female and a key part of the fertility rites that formed a widespread native religion.

*THE TERM 'GROUNDING' refers to any technique of breathwork or relaxation that brings a sense of inner peace. There are many techniques that help turn the mind from fear and excessive worry when faced with a stressful situation. Mostly they are used to quiet the mind before a meditation and, with practice, become a way to stop the rush of thoughts that hinder meditation.*

# The Great Circle at Callanish

I first visited Callanish on the night of the full moon in August. Looking to the land as my guide I walked outside the circle until I felt I had identified the entrance. Here it is possible to see the processional route into the circle, as there are four distinct avenues that go to the centre, where there are a cist and a tomb. The stone that towers above all the others stands at the centre of the circle. In fact the area the stones enclose is not exactly round; it is a flattened circle with the spaces between stones differing in size.

I found myself alone at the centre and lay gazing up at the towering stones. I wanted to curl up into the foetal position but there was another visitor on the edge of the site and as a woman on my own in the dark I wanted to see if anyone was coming my way. I always want to see stones at night, but sometimes if I do not know the area I encounter deep-seated fears and it was in Callanish that I met my daemons.

Callanish is a lunar-aligned site, so I wanted to see where the full moon would rise amongst the stones. The second night that I went there I was able to spend significant time meditating and had the pleasure of watching the moon rise majestically through a space between two of the entrance stones. The effect of the four distinct entranceways is that the circle is divided into quarters and by standing in the centre you experience a vortex of energy coming into the central stone. I actually felt the energy surging upwards into me as I spent three hours at the central stone.

I had decided to watch the moon rise from low on the horizon to a place above the stones, accompanying its movement by playing a Tibetan bowl.

**THE CHAKRA SYSTEM** *is part of Hindu philosophy and refers to at least seven energy centres in the body, each of which is responsible for certain organs and parts of the body. Each chakra is also associated with a different colour.*

**base chakra** *(between the genitals and anus) –* **red** *governs survival issues and the bowels*

**belly – orange** *governs the creative aspect and sexuality*

**solar plexus – golden yellow** *governs the sense of personal power and dominion*

**heart – green** *governs all issues of trust and letting love in and out*

**throat – turquoise** *governs communication and expression*

**third eye – violet** *governs the psychic sense and intuition*

**crown – white** *governs the relationship with a higher source or deity*

*In the West we have little information about ancient knowledge of energy centres but we do use visualisation to achieve the same goal – that of stilling the mind. It is possible therefore to address each chakra and visualise that particular energy growing or balancing.*

As I played I heard the sound of intoning and stopped to listen. The stones in the night have a luminous quality and as I watched and listened I was sure they were moving. At times it seemed as if there were people passing between the spaces in the stones. I had to continue to breathe to keep my nerve.

The stones seemed like weird sisters in the night. They are made of Hebridean gneiss, a stone that has a multitude of grains and shapes intertwined in its fabric. I watched the moon rise, playing all the time, and the sound of the bowl surrounded me with its eerie harmonics. I watched the moon rise through the layers of black and bluish clouds, hazy and slow, and, standing alongside my awesome silent sisters, I felt the fear I sometimes feel in the dark. Like priestesses we stood and I played and kept a wary eye on the space between the stones. At times it seemed as if I would give in to my fear, but I worked hard to stay still. It made no sense that I was seeing movement, but it gave me enough of a scare that I had to use the energy of fear to propel myself out of my desire to run.

Using my knowledge of grounding and stilling my energy, I sent a thread down from my base chakra deep into the earth. I focused the thread with my breathing and imagined it moving through all the layers of earth, through rock and porous roots until it reached the centre of the Earth. Once it made contact with the centre I pulled up the energy of the core through the thread. This is a technique I learned from Vivianne and Chris Crowley when I was first learning about Wicca. The thread keeps you grounded while energy is pulled up through it, giving life to each chakra. As the energy rose I drew it in and made each chakra glow with the colour associated with it.

## TWILIGHT WITH THE WEIRD SISTERS

*Strangely, I am normally more afraid of the city at night – having been raised in the country I trust it – but sometimes I am completely caught off guard by my fear of the dark and the unknown, and it requires technique to deal with it. In the centre of Callanish I would not give in to my city-girl fears. I held the moon between my fingers, focused and actively breathed out fear. The sound I used to release it was a breathy sigh and it took several breaths to regain my sense of confidence.*

**ONE OF THE SMALLER CIRCLES** at *Callanish. To me, one of the fascinating aspects of the Isle of Lewis is that so many sacred sites – including a total of six stone circles – are to be found within such a small area.*

Once the energy had risen to the crown chakra it flowed down in silvery light around my body into the Earth. Energy like this is a renewable source coming from deep within the Earth, then rising and returning to the Earth. It is contained in this cyclical system so that none is lost.

As I reached the crown chakra I captured the moon between my fingers which I had placed together to form a triangle. There I held it in my gaze and asked some deep questions about the nature of my fear.

Just about that time a couple arrived at the stones and though I greeted them with a 'hello' they did not answer, nor did they enter the central part of the circle. Twice I said something to them, but they never spoke. Later as the moon rose higher through black streaks of cloud, someone came into the circle with me. I could not discern their features, nor could I tell if it was a man or a woman, but he or she was accompanied by two dogs. I started playing the bowl again and one of the dogs came and sat between my legs for a very long time. Again I said something to this dark presence, but again nobody replied. It is well known that the country code is to say hello to people who are walking, and that in circumstances such as encountering strangers at night, a woman saying hello needs a response to reassure her of her safety. This night nobody spoke, and it was no coincidence that I was actively feeling fear and exorcising it. The silence was part of the experience. I had to tell myself repeatedly that everything was all right. The dog at my feet brought the feeling of safety and it was then that I felt myself slipping into a trance.

All my fear and attempts to deal with it resulted in a rush of upward moving energy. I felt high and ethereal, released and free of my fears. At first I thought it was the result of my meditation and my efforts to control the fear I was feeling. But then I recalled what Margaret had told me about Callanish being the sending place of the spirits. Perhaps the otherworldly light-headedness I experienced was the trace of previous ceremonies.

## *The Smaller Circles of Callanish*

One evening I saw a group of standing stones, black and distinctive, against the skyline and felt drawn to them. It was a beautiful site and very private, so I asked the entrance to make itself known to me and followed my instincts to a stone on the far side.

I wanted do an Earth rite, so I took a piece of the peat that covers Lewis and is used as fuel in the old black houses. Peat is an element of both Earth and Fire, suggesting land and life-force energy. I had come with the intention of 'grounding' and what better way than to take my clothes off and sit naked on the peat in a stone circle? I was reminded of the old ways of American Indian women who, when menstruating, would retire to the women's house and bleed into the moss. This is not to be underestimated as a healing act. As part of my centring, I wanted to feel the earth beneath my bare feet and naked body.

As I played my Tibetan bowl, naked and earthed, I felt that this circle was a place of joy. As darkness descended, I began to sing my heart out, using any lyrics that came to mind. The words described the land of which I was part and the loneliness women must have felt through the ages. This is the sense of place in the universe that you perceive when you look at the night sky and realise you are so very small. Yet in this knowledge is the wisdom that we are all interconnected by the sky and the stars and the earth we walk upon.

Though a visitor to Callanish, I have a deep connection to wild empty places and this circle felt safe and intimate. After playing the bowl I made love to myself and felt a sense of happiness that there are still places where the temple of the Earth and the naked human body are divine. The religion of the Hebrides is one of the most reserved and rigid forms of Protestantism known. Here nice women do not flaunt their sexuality or do anything to 'offend' their male God. I, on the other hand, feel the sense of life force rushing to me in wild lonely places and it is a source of potent sexual awareness.

# MAID, MOTHER AND CRONE

The seasons of womanhood are linked to Celtic spirituality by the archetype of the Triple Form Goddess. The Maid, Mother and Crone are equal in their potency. None can exist without the others and their teachings constitute a reverence for life in all its stages. Though the representation is always of wise women, the seasons of human existence also describe the masculine experience. It may just be that the definitive move from young girl to a woman who menstruates and is able to conceive a child to the cessation of the flow of blood at menopause is so obvious that this process is more available as a female archetype.

Though the Celtic cultural influence came much later to these islands, there is a link between their spiritual ideas and those of the earlier Neolithic peoples. This theme is also continued in the modern beliefs and practices of Wicca. The Triple Form Goddess is historically seen as the three Muses, the Weird Sisters of Shakespeare's *Macbeth* and in numerous statues and paintings throughout Europe. Within Celtic folklore and the oral traditions later written down as heroic mythology, she recurs frequently.

The Irish Goddess Bride (see page 99) is a case in point. She governs three distinct categories of 'power': poetry, smithcraft and healing. Yet she is always a Fire Goddess. It has been suggested that there were perhaps three Brides who governed different aspects of Fire. Fire is the element of inspiration which brings forth poetry. It is also needed to melt metal for smithcraft, and it is the fire of human warmth that heals through love.

**GEOMETRIC FORMS** *at Callanish (previous page and opposite). The stones at sacred circles are often distinctively shaped. At both Callanish, as at Avebury, diamonds and phallic shapes may be seen.*

The Triple Form Goddess is seen at all times as the interplay of three in one. Though the ageing process brings its changes, there is a little of each stage in each of us at any age. The Goddess in Wicca is a Goddess of inevitable and necessary change. At any given moment she can have qualities that make her a Maid, Mother or Wise One, as the Crone is often more aptly named.

In Wicca, the three stages of womanhood are all given honour. The Maid is not a virgin but a woman who is independent and true unto herself. She is able to take lovers but chooses not to take a spouse, as her independence and power come from individual liberty. She is a temptress and a woman who loves her freedom. She is the laughing, playful role of woman who relies on her own skill and instincts to move through life.

The Mother is the blood of life and the nurturer. She has chosen to give life and her role in the trilogy is a powerful example of compassion. She is also the guardian of the young and the old. Rather than the placid Christian image of 'Mother Mary', she is fierce in her role as protector.

The Crone is the keeper of the secrets of 'woman magic'. She has the wisdom of years and experience that is gained through time. Both in ancient times and in modern Wicca she has the status of an Elder, yet she is outside any system and can be wild like Hecate, the archetypal Crone Witch, uncontrollable and passionate. The wilder side of the Crone comes from her lightness of being. After a long life, the seriousness with which younger women take themselves becomes too ridiculous. She has the ability to laugh at herself and at authority because she *is* authority. Her decisions are informed and wise. She is the role model for all and her wisdom comes through the understanding of patience.

All three have healing abilities. The Maid represents youthful freshness on the journey through life. The Mother symbolises compassion and fierce protective magic. The Wise One has the wisdom of the ages. Yet within each stage there are the three aspects. Thus the Maid passes through three stages of Maidenhood, the Mother through three stages of Motherhood and the Crone through three stages of Cronedom. A Crone entering her youthful Crone period may be just adjusting to the time when the blood stops flowing out of her body and instead stays within to feed her, whereas the later stages take her into a calmer period of wisdom and ultimately to old age.

There may be times in life when one of these forms is useful as an image in meditation or in the rites of passage, for each also has the three in the one. The young maid has a fresh approach to life and is not encumbered by fear; thus she may also have intuition that speaks to her heart, that is Crone wisdom. The Crone equally has lost the inhibitions of the young and can be true to herself. She cares little what people think of her and is unencumbered by society's rules and regulations. The Mother is both youthful and wise. She has learned the selflessness that comes with unconditional love and once her young are grown she will rekindle the playfulness that her children taught her.

## ᙏaid, Mother and Crone at Callanish

Within one of the small circles that can be seen from the major one at Callanish stand the three stones known as the Maid, Mother and Crone. There is also the male consort stone, that looks remarkably phallic. He shimmers with white quartz, which runs down the stem of the upright stone like glistening semen. The Maid is triangular and white and the colour reflects the Wiccan symology of the maiden as pure and white. Again this means not untouched, but youthful. She represents the right to choose lovers without being tied to one.

The Mother stone stands between the others. She is red like the blood of the birth waters and of menstruation. The Crone stone is tall and her veins are intermingled white and black, causing the overall colour to be grey. Like the stones of the central circle, she is made of Hebridean gneiss.

Each of these stones has been carefully chosen and local mythology tells us that they have been known as Maid, Mother and Crone for a long time. There is no written text to suggest why, but it is interesting that the colours correspond to the modern Wiccan symology of the Triple Form Goddess.

Looking at them from the position of the phallic consort, the sisters sit in the foreground of a magnificent view over the Sleeping Beauty, the sacred landscape of the Goddess. To the base of the circle that contains the Maid, Mother and Crone stones, there is a sacred well. Unlike the wells in Eire which are actively used (see page 97), the beautiful spring at Callanish is not

**THE HIDDEN SPRING** *of the Callanish Goddess. It is interesting that in this sternly Protestant country the wells are neglected, whereas in predominantly Catholic Eire they are still used and one sees a mixture of Pagan and Christian offerings at the same site.*

marked. The link between the Goddess and the spring of life is to be found in many sacred landscapes. Yet this important Callanish spring has been neglected and is not even on the map in terms of its relationship to the stone circles. The spring is the shape of a kuna or cunt. (It is worth mentioning that I use these words deliberately. I shall talk more about the ancient and sacred names for the female genitals in the chapter on Avebury – see page 130.)

I found the well down a small embankment, clearly built into a shrine with an overlap of Pagan and Christian imagery. The spring rises on a line between a series of smaller circles and the main Callanish site. Since it lies so close to the Maid, Mother and Crone it is possible that this is a remembrance of a major Goddess site.

Why this spring is so neglected is debatable, but in the course of this journey I have visited many sacred springs that have been taken over by Christian saints. I gleaned clues as to why some were active and some not from observing the attitudes of the locals. What is interesting is that even if sacred springs are known to be Pagan, in Catholic areas they are still used. Catholicism's reverence for the Mother principle is actively contained within its religious teachings. Mary, the original great Goddess, is a co-creator in the mythology of the Catholic Church. In fact Mary is one of the most accessible saints whom devotees can ask for help.

I have no way of assessing whether this plays a part in the continuance of well rites in Eire as compared to the distinct lack of blessing at wells in predominantly Protestant Callanish. But I can say that in Eire I saw many local people driving up and spending a little time at the wells. They often drank the water and lit a candle. In many cases they did this without any clear understanding of the history – it is simply believed to be 'good luck' or 'healthy' to drink the waters. Leaving a rosary or a note to ask for help exactly mirrors the actions of Pagan ancestors leaving offerings for the Goddess. Is it the connection to the Mother source – strong in the Catholic Church but neglected in the Protestant – that allows these ancient customs to persevere?

# Rites of Passage: personal rite for the three stages of womanhood

Depending on the aspect with which you associate most closely, it is possible to use the model of Maid, Mother and Crone to do some personal work. What is important is to realise that the three are one and that if we are to understand the cyclical nature of life, the teachings of each stage are equally important. Men and women who have not had children are not exempt from the teachings of the Mother. Young men and women have access to the wisdom of the ages and are indeed wise. A Crone may have the playfulness and the energy of one who doesn't give a damn while also understanding the significance of the later years. They are all powerful personas for anyone to contact and to connect with.

**CRONE, MAID AND MOTHER STONES**
*(opposite and on pages 73 and 74). In ancient religions woman was revered as the giver of life and in modern Wicca the three stages of womanhood are equally valued as part of the cycle of life.*

## CRONE RITES

*While many sacred rites traditionally take place at the time of the full moon, there is much teaching regarding the period each month when the moon is on the wane. The dark of the moon, as it is called, is the time when the energy of the moon meets the dark aspect of the Goddess. This dark Goddess is she who turns inwards and accepts all aspects of herself, even those characteristics that seem less than desirable. She is the wisdom of the shadow; that which is deeper than all that is seen, and she takes you to depths that cannot be felt when 'light' is shed upon the heart.*

*'Working the dark' is most appropriate when there is something to be uncovered, something that may not sit easily with the spirit or that needs to be released. In the course of a cycle of the moon, a Witch may decide to work on a particular issue both at the dark and at the full. On the dark she or he may work to remove the obstacles that prevent the fulfilment of a desire and then on the full, try to bring the desire to fruition. When working with an illness, for example, it can be useful*

*to look at attitudes that you wish to let go of to aid recovery (the dark) and then work for the improvement of health (the full). Anyone who wants to let go of a lover, a job or a painful situation may also work at the dark.*

*Any issue pertaining to the Crone may be addressed at the dark of the moon, though this is not the only time to speak to her. You may want to honour a certain*

*person who is about to turn a grand old age. For this special time of life, usually over 60, there may be a way to take them to a sacred site and do something that marks the occasion. Dolmens and cairns, where the ancestors stored special bones of the dead, are perfect for Crone rites. It is important to realise that the Crone manifests on the path of men, too. She is the aspect that many people of both genders look to when walking the path of age.*

## CORD RITE FOR THE CRONE

*Take a black, blue or purple cord and tie intention into it, as described on page 35. Speak the spell into the knots, then cut the cord and leave it. If you are doing this in a sacred space, it may be possible to find a nook in the rocks or in the standing stones themselves where you can leave the cord. Alternatively you can bury it. This is a way of cutting through a difficult situation and leaving it behind you.*

*If you are honouring a Crone, she or he may want to keep the cord. In this case it becomes more of a wishing cord for someone you want to honour and bless. There is no need to cut the cord for a blessing.*

*The words or chants that are used over a cord or a candle vary, but when creating your own ritual say them three times, or three times three, to seal the working. This repetition honours a cycle of wisdom that passes through the gathering, maturing and manifesting period, which again makes reference to the Maid, Mother and Crone but also ties in to the cyclical nature of human experience and to the monthly moon cycle and seasonal changes.*

## BULB RITUAL

*In some cases the Crone is an aspect of yourself that you wish to address. This can happen at any age, particularly when a relationship has just ended. Often those are the times when we are forced to look the Crone in the eye, to see who we are as individuals and not in the safety of a relationship. It is the time when we feel the meaning of the words, 'You come into the world alone and you die alone.' As well as the vulnerability there is a sense that we are what we are, no longer dependent on another to define our tastes, our personality, our pleasures and our needs. When the love is lost it is time to love yourself.*

*Sometimes it is useful to be able to leave parts of yourself behind or to cut through a situation. For this you may wish to perform a bulb ritual. It is best to do this in the autumn, for when you bury the bulb you can look forward to seeing it bloom in the spring. It can be done at other times, too.*

*I have found bulb rituals effective at times when I really needed to go underground and seek solace from the quiet earth. In some of my rites I do the work at my altar and then take the bulb to a special place to plant it. For the Crone I choose a bulb from the gardening centre near me and place it on my altar. Then, using incense to create sacred space, I meditate with the bulb in my hands. I feel all the age, all the weariness, all the disappointment and the dashed hopes that I can possibly conjure. I actively send this energy into the bulb. I speak words that I would not wish anyone to hear, for they are my Crone thoughts. Uncensored I speak and cry and cackle like a mad old crone until I really feel discharged.*

*Send this energy, private and at the end silent, into the bulb. Send it out with love, knowing that the negative energy will change in the course of time. In the spring you are able to witness the change as the flowers raise their shoots and heads. After winter's respite this is the time of new growth; the transition from Crone to Maid is visible. If possible, bring some of the flowers into sacred space. If you want to bless a few bulbs, choose ones that flower at different times so that you can see them emerge all through the spring. Knowing that these flowers once harboured deep Crone thoughts and that you are now free of them, brings an immense lift to the spirits. Once the flowers have come up, perform a rite to acknowledge the Maid in you. Cut them, honour the cutting and leave the situation behind. This is a strong way to move through from one stage of personal magic to the next.*

## MAID RITES

*The Maid is the aspect that we look to with regard to anything new. It is the time to start new ventures or to use the energy of spring to prepare for a new cycle of life. Whereas the Crone is the cutter, the Maid is the flame of inspiration. The important thing about a Maid rite is to start anything you intend to do in a state of mind that conjures up the youthful aspect of your personality. Wear bright colours and colours that emulate nature in spring. Choose candles that are yellow, light blue, white or cream. If possible, visit a sacred site in the spring. There are numerous wells from which it is possible to take water in order to seal fresh intent into the candle. It may be hard to light a candle in some of the more remote and exposed stone circles, but it is always possible to carve a symbol of your intent into it while at the site and then light it on your altar later.*

*Meditate before carving the candle and allow yourself to go back to the essence of your youth that brings forth easy laughter, fearlessness and the ability to take a risk. If you feel too tied to the mundane in your life, light the candle and concentrate on the things you have always wanted to do and haven't. Write down these honest desires. The act of bringing them to the surface does not mean that you want to do all of them, but it carries with it the energy of the possibility. Pass the candle through incense while facing the east, the direction of the rising sun. Equally potent is to carve the candle as the moon rises and then to light it and let it burn for the entire night.*

## MOTHER RITES

*The Mother goes hand in hand with crystal magic. She is Mother Earth and her stones are her harvest. Crystals and stone circles have a special relationship, which is why crystals are so effective at these sites. Many circles have quartz crystals not only in the stones themselves, but also buried underneath the stones. When quartz is in its natural state, striking two pieces together creates sparks, which is why they were considered sun stones. The best examples are the smooth quartz found on beaches that have been rolled by the tides. These stones appear not clear but white. In any Mother rite they may be the preferred crystal, but many others may be used because of their particular qualities or colour.*

*Keep an eye out for quartz on the beaches near a sacred site. The sea yields much and it may be that you wish to create a small altar of these stones and then, as you*

*state your intent, spark the stones as an affirmation. There are many books written on the subject of crystal healing and the properties of individual crystals, but the Witch's way is to use the crystal that speaks to you. This is a form of 'scrying' – a special Witch's method of divination – that involves going into a shop and looking at the crystals while actively sending out a request for the one that suits your purposes. You may have the opportunity to close your eyes and pass your hand across the stones and you may start to sense one that is right for you. I was taught a long time ago to use my left hand for any sort of divining, including choosing a tarot card or rune, and I always do so. The sense that develops is like a slight buzz, tingle or heat sensation. Some crystals feel hot even though the stone itself is cold. The key is to trust yourself implicitly and choose the stone that won't go away.*

*I had an experience once that stays in my thoughts whenever I choose crystals. I had stopped in a crystal shop on the way to the Long Man of Wilmington in Sussex, a sacred place with the effigy of a huge chalk figure carved into the hillside. I had gone into the shop with a friend who wanted to buy, but I intended just to look. The fact was that I didn't have much money to spend on new stones. The shop, however, had recently had a shipment of fluorite, which has always been one of my favourites. It ranges in colour from deep greens to pinks and purples, and in this shop they had one of the most beautiful large green crystals that I had ever seen. Though there were many, this one was speaking to me. I held it and felt the familiar warm tingle. I looked at many other stones, but still came back to the same one. Nevertheless, when my friend had chosen the crystal she wanted, I put the green fluorite back on the shelf and left the shop.*

*The next stage of our journey was to walk up the hill of the Long Man, but from the moment I started the walk I couldn't think of anything except the stone I had left behind. I had a conversation in my mind about how I would see others just as nice when I had some money and that this wasn't necessary at the moment, etc, etc. This conversation continued all the way up the hill and all the way down. I got into the car, drove straight back to the crystal shop and bought the fluorite. I had ignored my heart and tried to justify it through financial considerations, but the stone would not allow me to do this.*

*It taught me a great lesson, for just as I have never seen a stone like this one, I have also seen crystals given back to a place or a person. In two significant rites I have accidentally left wonderful crystals at a sacred site without remembering until it was so late that I had to learn to let them go. I left a heart-shaped fluorite at the Long Man of Wilmington when I performed a handfast for two men. Same-sex marriages are usually condemned in orthodox religions and even in Wicca they are not as common as I would like to think is possible, but I had the honour of joining two dear men in a love ceremony, Wiccan style, and I had taken a very beautiful crystal heart as a symbol of love. It never came back with me, though I packed the things I had brought very carefully. I let it go back to the land as my offering.*

*It was a holed stone that taught me that it is sometimes permissible to take stones out of a ritual. I was working as an archaeologist in Sussex and came across the site of a Witches' or Pagan rite on Chanctonbury Hill, a place that is famous for its associations with Paganism, and particularly Witchcraft. To my amazement I found a little grove of trees adorned with hanging garlands and the Witches' sign of the pentacle. I realised that the necklaces that hung in the trees had been placed there with love. I took one down and saw a perfect holed stone. After some internal deliberation, I felt that it was part of my journey to take it. And I did. I have it to this day and ever since, I seem to have developed the ability to find holed stones wherever I go. Before this incident I could never find one however hard I looked. Now it's a case of glancing down and more or less seeing them lying at my feet.*

*Finding that stone and asking whether it was OK to take it from that place was part of my journey to becoming a Witch. It was a symbolic moment in my personal introduction to Witchcraft.*

*Once a crystal has been chosen the process of charging it makes it a personal magical tool. Wearing a crystal close to the body is one way of doing this; another is sleeping with it under your pillow. If the crystal is to be used in healing it is wise to cleanse it before use.*

*There are many ways of using crystals in healing, but one of the most effective is to hold a clear quartz point down above the injury and to send energy through the crystal and out via its point. Move the crystal gently back and forth above the wound while sending energy. Keep going until you feel the energy is received. It is important to receive as well as to give. If working on someone else, ask him or her to open their hearts to the healing before you start. There are also methods of detecting the direction in which to move. Again this is done by speaking psychically to the injured part and tuning in to what the energy is doing. If it feels as if there is a blockage, use the anti-clockwise direction to release; to contain energy and promote healing, move the crystal clockwise. Sometimes starting close to the skin and gradually moving further away dissipates the energy.*

*The most important part of a healing is that when you feel it is time to stop, you must seal the energy in by making a mental break from the work you are doing. Look away, change direction, take the crystal gradually further away and finally place your hands on the ground.*

*Another method is to wash your hands, then to clean the crystal itself. One of the Witches' methods is to imagine a great cauldron of energy bubbling inside your stomach or solar plexus. The energy may be imagined as a golden light which pours out of the cauldron, out through the arms, hands and fingers, and into the crystal. When this process is finished, complete the seal by placing a lid on the cauldron and making sure no energy is leaking out. This is an essential part of the process, to prevent the person performing a healing becoming drained of energy.*

**CLEANSING CRYSTALS** *I have been taught many ways of cleansing crystals. One of my earliest teachers was a Quebecois Witch who made healing tinctures by imbuing water with the aura of the crystal. Her advice was first to study the essences of individual crystals in books by experts, and then to use this method. She put the stone in distilled water in a clear glass bowl during the hours of the full sun, from about 10 am to 3 pm. Then she sealed the water in bottles. It would now be a tincture of whatever stone had been placed in the bowl.*

*Since becoming involved in Wise Craft I have heard many other methods of cleansing crystals and imbuing them with intent. Some Witches place the crystals, point down, in the earth overnight. Most mention the sun as a major aspect of the cleansing but in Wicca, where the moon plays an important role, I have been taught to leave crystals overnight at the time of the full moon so that the moonlight has the entire night to imbue them. I use this method to clear the space every full moon when I want to renew energy. No matter what kind of rite I am doing, this is my feng shui technique, Witches style, to keep the energy refreshed and clear. I put crystals in a glass bowl, cover them with water and add sea salt as the cleansing force. The moon imbues while the salt clears.*

*I also use water from the sea if it is near a site where I wish to work. The ocean is the great cleanser and has the salt element. Rivers and streams are also powerful cleansers – I hold the crystal in my hands and allow the water to wash over the stones while actively, intentionally, cleansing the crystals with my mind. I send forth my own energy of clearing, down my arms, through my fingers, through the stones and out.*

*Using sacred spring water is another a powerful way of cleansing both crystals and your own energy. If possible, take stones to the wells and sit them in the water. I always have some sacred spring water in the fridge as I visit sacred sites regularly and when I am at home I use sacred water in all magical workings. This has an added dimension, for the spring water has been imbued for thousands of years as one of the earliest forms of divine Mother essence in Celtic mythology.*

*The fact that water is one of the most effective cleansers in elemental magic means that tap water may be used if no natural source is available. Hold the crystal in cold running water for the duration of the cleansing. With practice this can be done in minutes. Intention is everything, helping to move old, negative energy. Cleansing crystals so that they can be charged is essential preparatory practice and focuses the mind. Once cleansed the crystal or stone is ready to be used.*

# 𝓜AGIC FROM THE SEA

Orkney is full of Pagan sites. Some of the most impressive passage tombs and stone circles are found in the far north of Scotland. The Ring of Broadger is the most recent of the three sites that are part of the sacred landscape of these remote islands. Within view from the ring itself are the remains of an earlier stone circle, known as the Stones of Steness, and across the loch, the even older Maes Howe passage tomb. Maes Howe has been dated to 3500 BC, the Stones of Steness to 3000 BC and the Ring of Broadger to 2500 BC. Like Callanish, these sites are situated beside water and are striking in their beauty.

The folklore of Orkney is told in tales that reflect the playful nature of the islanders. They are a soft-spoken people and have many stories to 'wile away' the long winter nights. One tale tells how the Ring of Broadger was created.

*One night a group of giants went to the field at Broadger for a dance. The fiddler struck up a reel and away they went. They all joined hands and danced in a great circle. Round and round they went, but they were enjoying themselves so much that they never thought of how quickly the night was passing. Suddenly the sun rose behind the hills and they were all turned to stone. There they remain to this day. The fiddler can be seen standing in the field next to the dancers, for he is now the Comet Stone.* [1]

*THE RING OF BROADGER* (previous page and below). Opposite, the Stones of Bro, Gotland, Sweden, said to be two hags who quarrelled on the way to church and were turned to stone. Many stones across Europe and throughout Britain are said to be people turned to stone.

This theme recurs throughout the British Isles: other stone circles are remembered as people, usually women, who are turned to stone. The fact that they seem to dance only at night is mentioned so often that this is likely to refer to ancient rites that took place under the stars. If you spend time in a circle at night you may have a feeling that the stones move, as I did at Callanish. Their shape suggests human form and may well have been carved intentionally. One of the stones between the Ring of Broadger and the Stones of Steness is reputed to drink from the loch on the stroke of midnight on Hogmanay (Scottish New Year's Eve). It is known as the Watch Stone, and on this night and this night alone, it comes to life, walks to the shore and drinks the water.

Another theme is that giants placed the stones there. This may be a folk rendering of the construction of the stone circles, as it seems a 'giant feat'.

I have had the opportunity to put up two stones in my life and it takes a lot of strength and a lot of people to raise even one small standing stone. Even with rollers and womanpower it is an exhausting task.

Orkney is surrounded by water and the sea plays a significant part in the life of the islanders. The ocean is a goddess known as the Mother of the Sea. Historically the islanders relied on fish for their livelihood and a benevolent goddess meant a good harvest. There is a story that tells of the seasonal change in the nature of the sea. In the summer the Mother of the Sea calms the waters and aids the fisherfolk in gathering the fruits of the ocean. In winter a monster called Teran causes storms and rough tides. The battle between the forces of summer and winter are named the Vore Tullye or spring struggle. Inevitably the Mother of the Sea fights against this monster and wins. She binds Teran in chains and sends him to the bottom of the sea. Throughout the summer she guards him and the summer seas are calm and fruitful. However, by the autumn the Mother is weak from her work of keeping the monster in chains and Teran forces himself free. The autumn is the time of the storms, known as the Gore Vellye or harvest destructive work, when Teran wreaks havoc and his winter reign begins. The Mother becomes land-bound until she regains enough strength to fight the spring struggle and bring the warm calm seas of summer once more.

The importance of the sea is also reflected in the legends of mermaids and fin folk, which form a large body of Orcadian folklore. The mermaids often represent an erotic form of woman whom mortal men find irresistible. In the early versions of these stories, clearly written from a Christian viewpoint, a mermaid is able to remove her tail, but this is improper, unwomanly behaviour. The story goes that the mermaid was the most beautiful of women. One day a great queen, 'perhaps Mother Eve herself', sees the mermaid sitting on a rock, combing her hair and looking at herself in a mirror. (In later stories, the comb and mirror become potent symbols of the mermaid's freedom.) The queen is shocked that the mermaid wears no clothes and sends her a dress. The mermaid refuses to wear it and the queen is filled with jealousy. She gets all the women of the land to create a great fuss over the mermaid and to say that

it is a sin for so beautiful a creature to sit naked on the shore. The women are also worried that, with such a lovely woman in their midst, the men will have no eyes for anyone but her. The mermaid is accused of using magic to create her beauty and a spell for her seductive music. The women demand she be doomed to wear a fish's tail, which she can remove only if she takes a human husband.

**LOOKING ACROSS THE LOCH** *from the Ring of Broadger (main picture), dated to about 2500 BC. The Stones of Steness (inset) are 500 years older.*

Another tale, *The Mermaid Bride*, gives an idea of what happens when a Pagan mermaid interacts with the pious Christian community of Orkney.

*Johnnie, who is known as the handsomest man in the Orkney Isles, sees a mermaid on the shore. He feels the hot fires of love burning strong in his heart and swears on a Meur-steen (a stone on which oaths were sworn) that he will win her for his wife. Before she can flee he sneaks up and kisses her, but though she flings him to the ground with her strong tail and returns to the sea, in her haste she leaves behind her precious golden comb. She pleads for it to be returned, but Johnnie says he will not give it back unless she becomes his wife.*

*Johnnie's mother tells him that as long as he has the comb he has power over the mermaid. He locks it up in a safe place and waits. One night he cannot sleep and he thinks he can hear music. Drifting in and out of sleep he finds himself spellbound and sees the mermaid Gem-de-lovely at the foot of his bed. He tries to say a prayer but the words do not come. She asks if he will give back the comb and if he will come with her into the sea. He refuses, and when all her efforts to persuade him fail she makes him this offer:*

**I'll be your marrow. I'll live with you here for seven years, if you'll swear to come with me, and all that's mine, to see my own folk at the end of seven years.** [2]

*Johnnie swears on the Meur-steen to keep his part of the bargain and they fall into each other's arms. They are married shortly after and, as the priest prays, Gem-de-lovely stuffs her hair into her ears to block out the sound, for the sea folk cannot stand to hear the word of God.*

*She is the best wife any man could hope to have, but she always has a faraway look in her eyes. After seven years the couple and their seven children prepare to visit her people, the fin folk. The youngest child is sleeping at Johnnie's mother's house when Grannie takes matters into her own hands. She builds a great fire in her hearth and forms a piece of wire into the shape of the cross. At midnight she puts it on the fire until it glows red, then brands the child with the cross of God. He screams 'like a demon'.*

*The next day the boat is ready and Gem-de-lovely's people have come to help. The couple and six of their children board the boat. Some servants are sent to fetch the youngest child, but they return empty-handed. When Gem goes to Grannie's house she tries to lift the cradle and finds that it will not move. When she holds her child, a burning pain shoots up her arm and she jumps back with a scream. She has to leave her youngest child and, with tears streaming down her face, goes to the shore and sails away with the rest of her family. Grannie looks on from a rock with a 'laugh hanging around her mouth'. Johnnie and his mermaid bride are never seen again.*

*Grannie brings up the youngest boy, who is named Croy of the Cross. He grows up extraordinarily strong and, when his grandmother dies, he takes up the sword and goes to fight in the Holy Land. He slays many chiefs and gathers a great store of wealth before marrying a jarl's daughter and living happily ever after.*

*A LATE BRONZE AGE stone site in the form of a ship on the island of Gotland, Sweden. Ships would, of course, have been important to island people and this may have been a monument designed to appease the gods of the sea or to ensure a safe return for the sailors or fishermen of the community.*

What is amazing to me about this story is that the mermaid is seen as the antagonist. She is not of the God-fearing folk, so she is expected to adopt their ways. The grandmother is seen as a good woman even though she actually burns the Christian cross into the back of a small child. These themes are repeated in many Orcadian tales, and it is always the 'foreign' woman, the free-thinking sexual woman, the naked woman who is the evil character.

There is another kind of ocean-dwelling people native to Orkney, the Selkies. Rather than having fish-like tails, the Selkies look like seals with removable skins. Like the mermaid, the Selkie female is renowned for her beauty and she is often seen on the shores, naked and combing her long flowing hair. Selkies dance in human form at high tide, which is perhaps another reference to the rites of the full moon, when tides are high. This story concerns a group of Selkies who are seen on a tiny island that rises from the sea.

*One summer evening a man called Harold is out fishing when he hears the most wonderful music. Captivated, he turns his boat to follow the sound. He arrives at the island of Boray where he witnesses a group of people dancing, yet he cannot see any musicians and believes the music to be created by magic. Moving closer to*

*shore, he sees a pile of black objects that look like beasts and goes ashore to examine one. It is a sealskin. Watching the dancers spin round and round, Harold drops the skin into his boat and heads out to sea. As the sun rises over the horizon the music suddenly stops and the dancers all hurry to shore. They take up their skins, jump into the sea and swim off – all except one woman who is searching for her skin. When Harold goes back to give the skin to the woman he realises to his horror that she is his own dead mother, who drowned many years before. She tells Harold that it is the fate of all drowned people to become seals, but that on one night every month they can dance from sunset to sunrise in human form.*

*Harold's mother makes an agreement with him that if he gives her skin back she will promise him the 'bonniest lass among all the Selkie folk as his wife'. He is told to return one month later when she will show him the young woman's skin. The girl will be in his power as long as he keeps the skin locked away.*

*One month later Harold returns. His mother meets him on the beach and gives him a sealskin which he takes back to his house before sunrise. Once it is locked away, he returns to the island. When the sun comes up the dancers put on their skins and swim away – all except one beautiful woman. She wrings her hands, crying that her skin has gone and that she cannot join her own people. She tells Harold that she is the daughter of a Pagan king. He tries to comfort her and persuades her to return to his house and become his bride. They have several children but after a while the Selkie falls sick. She seems to waste away with some secret grief*

**MAES HOWE PASSAGE TOMB** Dated to 3500BC, Maes Howe is the oldest of Orkney's sacred sites and contemporary with Newgrange in Eire. It is aligned so that it receives the rays of the sun on the morning of the Winter Solstice.

and begs for her sealskin. Harold refuses because he cannot bear to lose her. Then she confesses that she is worried about her soul because she is a Pagan. A priest is called and she is baptised a Christian, but it does not stop her from pining away.

'Harold,' she says one day, 'we have lived long and happily together. If we part, we part forever. If I die I cannot be sure that my soul is saved, for I have long lived a Pagan. Tonight is the dancing night; roll me in my sealskin and leave me on the beach. They cannot take me away if I am Christian. But you must go out of sight, and return for me in the morning; then you will know my fate.'

Harold does as she bids – finally – and sits waiting all Midsummer night. The silence is broken by a sudden wail. The Selkies have found his wife on the shore and are lamenting that they cannot take her with them. When the dawn breaks, he finds her on the shore, dead.

**On her sweet face she wore a smile, the sign of a soul at peace. That smile comforted Harold, for now he knew their parting would not be forever.** [3]

The moral of the story is clear: the death of a Pagan does not matter as long as she is saved by her conversion to Christianity. Her death is not the end because Harold knows she has accepted the true faith.

When I read stories like this I can imagine the way they were first told. The Pagans are the problem and it is of the highest importance to protect the protagonist from the influence of another culture. In some cases it is as blatant as in the story of the Selkie of Boray. She is a Pagan and the daughter of a Pagan king. Harold has to convert her and force her to live a married life with him. Yet these actions are apparently governed by love. The rereading of these stories is an essential part of Pagan history. In all of them, women as powerful sexual creators are forced into some form of submission, compelled to choose between their children, their families and their freedom, in return for being married and tied to a God-fearing community.

*THE FAERIES live in mounds, usually the sites of Neolithic chambered tombs or Iron Age brochs (circular towers). Faeries and trows (trolls) have the reputation of shooting at inquisitive people or cattle who get too near to their homes. They are experts at firing 'elf shot', which can cause great pain or even death. Neolithic arrowheads are thought to have been this magical weapon. In fact Neolithic flint arrowheads were kept as a charm as they were believed to protect the owner from harm.*

Tom Muir, who has collected the stories of Orkney and the smaller islands that surround it, describes several other-worldly creatures that may have roots in the ancient lost lore of Britain. They are shrouded in mythology but some glimpses of pre-Christian customs are seen in these tales.

Faeries loved music and dance. Often the tale is told of a mortal entering a faery mound and being invited to a great feast. The evening is full of revelry and the mortal has a wonderful time but, on leaving, discovers that what seemed to be one evening has actually taken much longer. In many cases it is the magical number seven years. The tales consistently mention Celtic festivals as times when the faeries are most powerful: Halloween, Midsummer and Midwinter are the times when mortals must take great care.

In rereading these tales it is possible to see the faeries as the older peoples who descended from the Neolithic era and maintained the old customs. The 'mortals' are Christians, always warned to keep away from the mounds and not to partake of the faery draught. This infamous potion was possibly a form of the hallucinogenic drink used in sacred rites the world over. Whatever it was, it created a desire in the mortal to stay and join in with the revelry. Faery folk were extremely attractive and there is a suggestion that they were open with their sexuality. It seems likely that these earlier cults did not fear their God/desses, but loved the freedom of expression they permitted.

This worried the tellers of the tales, who always approvingly describe the Christians as 'God-fearing'. They warn against drinking the faery draught (which may be interpreted as straying into the realms of ancient spirituality) and having too good a time.

THE MIRROR has a secondary function in the type of scrying described here, for it also equates the ability to see into the patterns of the universe with the courage to look into the reflection of the mirror. The old adage 'Know thyself' refers to the ability to look deep within and release false motivations. It is a divining tool of truth in that it requires knowledge of your own soul so that you can impart helpful information to others while remaining in the position of 'witness'. This is not meant as a way for you to influence another's thoughts, but to allow the flow of energy to move easily into and out of your consciousness.

# The Mermaid in Cornwall

The ocean has a mythic force in Cornwall, too. The people who have for centuries relied on the sea for food and transport looked to the Goddess of the Ocean to bring forth the fruit of the sea. In the church in the town of Zennor there is a fabulous wood carving of a mermaid holding a mirror and a comb. The mirror is interesting, for it is one of the most potent tools of scrying or divination in the Witch rites of Britain. The most common form of scrying mirror is made of shiny black glass and is often convex so that it can sit on the ground or a table and appear like a pool of black water. Like the crystal ball, it requires the diviner to soften his/her gaze and stare into it as a stream of thoughts passes through his/her mind. The key to both these methods of divination is the ability to lose one's thoughts in the process and so gain access to other information.

The comb which is the pivotal point in the story of the Mermaid Bride in Orkney also appears here. In Cornish folklore there is a suggestion that the comb was not a comb but a stringed instrument. In most stories the mermaids have a special affinity with enchanted music. The comb may have originally been a Celtic harp, an instrument which was certainly used by the Druids as part of the sacred training of the Bard. In fact the music of the sea is supposed to drive sailors mad with desire – they sail boats too close to the rocks, wrecking them in their attempts to get to the beautiful sea creature.

The enigma of the ocean is paralleled by that of the mermaid. According to one authority, Ian McNeil Cooke, the 'mer' does not originate from the French *mer* meaning

*THE MERMAID AT ZENNOR sits in a church holding the ancient emblems of a mirror and a comb/harp. Many stories of mermaids tell of them using enchanted music to encourage men to follow their hearts rather than their minds.*

'sea', but from 'meremaid' or 'merrie maid'. The word merry, which is used time and again in Witch lore, is the key to the role of the mermaid in Goddess worship. She is akin to the goddess Aphrodite in Greek mythology and is also connected to the moon. As the Goddess of love she teaches enchantment and following the heart, not the mind. Like the Greek Sirens, her sexuality kills men.

The Cornish legend of Luty and the Meremaid describes the lure of the woman as a sexually desirable creature who knows the secrets of the sea. The God-fearing Luty is enchanted by a beautiful sea woman known as Morvenna. Her underwater realm is a place of death to the landlocked Luty. Morvenna invites him to the place where *... the noble sons and fair daughters of the earth, whom the wind and the waves send in foundered ships to our abodes .. ( and their ) bodies move gracefully to and fro with the motion of the waves, (so that) you will think they still live.*[4]

Sexuality equals death to Luty. The meremaid inhabits a realm dangerous to man. She lives by her own rules and, though men are compelled to follow her, they will die in the process. This said, many do want to follow or force the merfolk to come to land.

The ocean has long been seen as a place of strong sexual urges and emotion. In Wicca, the element of Water is one of many mysteries and the greatest mystery is the container of the water – the cup, cauldron or grail. Men who seek the grail meet many perils en route and some never return. The greatest mystery of all is Life, and Life comes from the watery realms of woman. The wetness and the blood of life are both sacred and dangerous in misogynist Christian interpretations.

# St Nectan's Kieve, Cornwall

This site is strongly connected to the symbol of water, and is found at the top of a beautiful walk up St Nectan's Glen. A hermitage building sits atop the rugged cliff over which the stream drops, creating a magnificent waterfall. Kieve is the Cornish word for basin and the Kieve is created by the thundering water that has for centuries etched a basin into the rock beneath. It is a site full of female mystery, for the water rushes through a round hole at the base of the cliff and again the imagery is that of the source of life, the kuna or pussy of the Goddess.

The Kieve is found by walking up a ravine to the source of the fresh water. Myths about the Kieve mention it as the place where the Knights of the Round Table started their quest for the Holy Grail. The symbology of this story is that from a source of water, a place of Goddess veneration, the knights set off to find the ultimate symbol of the Goddess – the grail.

The steps up the cliff were cut out of solid rock and have been the path for pilgrims for 1500 years. St Nectan was a 6th-century Celtic Christian who lived as a hermit above the falls, but his patronage of the site took over from the earlier Celtic Goddess lineage. After his death two women, who may have been his sisters, diverted the river and buried his body in an oak chest under the falls. The river now runs over the burial place, in accordance with St Nectan's wishes.

**MAZE CARVINGS** *near St Nectan's Kieve (above) are reputed to be 5000 years old. The stream that flows down the ravine reaches the coast near Tintagel Castle (opposite), legendary birthplace of King Arthur and his half-sister Morgan-le-Fay.*

Not far from the Kieve are some of the oldest known carvings of a maze, identical to those found at Knossos in Crete. The use of mazes in meditation was widespread in Britain and there are numerous grass and hedge mazes that have been used by monks and Pagans alike. Following the path of the maze to its centre is a way of concentrating the mind. It is also possible to use your finger to trace the outline while meditating. Again, the important thing is to clear your mind and to focus on your intention.

## RITE AT ST NECTAN'S KIEVE

*I conducted a simple rite here for a group of American women. One had found out that morning that a dear friend had just died and she was in a state of shock. We walked along the glen in silence and set our minds on what we wished to do at the falls. The steps up to the hermitage enable you to stand above the falls first and then to descend. This is ideal for doing a simple rite in three stages. The walk itself is long and meditative and is an ideal way of focusing the mind. It offers the perfect symbolism of life's path, for along the way a beautiful stream meanders beside you and then you cross it by means of a little bridge. At the end of the path are some steps up. Once you reach the steps, the work begins.*

*At the upper level of the falls it is possible to walk out and look right over the drop of the rushing water. Here we all lit incense and cleansed ourselves with it. Then each of us spent some time looking over the falls and letting go of any worries. The place enables you to be almost 'in' the rushing water and the sound immerses your being with the image of cleansing. We let go, and the woman who had lost a friend dropped some flowers over the edge one at a time, saying goodbye in her own way to the man who had died. Some of us who had started walking down to the bottom of the falls saw the flowers floating through the opening in the rock. The image was one of beauty and hope. This simple act of saying goodbye into the water felt very meaningful.*

*At the bottom we held hands and danced together. Some of us waded into the water and blessed a chalice, the symbol of love. As we raised it, the image of the knights toasting to their quest was almost palpable.*

## MIRROR WORK AT THE DARK OF THE MOON

*The mirror referred to in the mermaid stories may well have been one of the magical tools of our Pagan heritage. Certainly the black mirror is used in Witchcraft as a tool of divination. It is the gateway to the soul and to looking within. The scrying tool of the mirror is most potent at the dark of the moon.*

*I have addressed many uncomfortable situations in my life, mostly in myself, and these obstacles, though painful, I understand are necessary for my growth. I do not wish to sweep them under the carpet but to gain a greater understanding of my own patterns or misconceptions, and for this I use a black mirror.*

*When I was in Orkney, the place of the mermaid and her tool the mirror, I had occasion to use the dark of the moon to look at something very uncomfortable. A dear friend of mine had died and left a mystery at the time of her death. There were many questions to be answered and since she was not available, I wanted to look deeply at the mystery without referring to anyone else. I had been unable to attend her funeral and the sense of incompleteness was unbearable. I knew from two previous visits that the Ring of Broadger was an extraordinary place, so on this dark of the moon I went with my friend Michael to a place of ritual to try to achieve some peace.*

**RITUAL AT ST NECTAN'S KIEVE** – the piper calls. It was from here that the Knights of the Round Table set off on their quest for the Holy Grail. From a place of Goddess veneration, they went in search of the ultimate symbol of the Goddess, the cup or Grail.

*I consecrated a circle within the stone circle. I created sacred space with incense and went to the west, the place of death and the place that Wicca ascribes to the Feminine. Once we had raised some energy by spinning round and round in a widdershins direction we set about the work I wished to do.*

*It is usual in Wicca to work deosil or sunwise. This is the clockwise direction and it focuses energy to build a rite. Moving in a widdershins or moonwise direction, anti-clockwise, releases energy, and it is the energy of the Crone. On the dark of the moon and on Samhain (Halloween) it is traditional to work widdershins. Samhain is the feast of death and the time when the energy of the year is released. It is also the Celtic New Year and the time to celebrate with those who have passed over. It is not a morose ritual but one that invites all those who are dead and loved to attend.*

**THE RING OF BROADGER BY MOONLIGHT** was a perfect place to work on a problem to which I had been unable to find a solution. Thanks to my scrying mirror I was able to accept what had happened and achieve a sort of peace.

**SCRYING** offers a sense of completion in situations where there can be nothing but the bigger picture. Ultimately I believe that the universe offers solutions to all problems, no matter how bleak they seem. Sometimes we cannot see why something has happened, but much later there will be some form of resolution. Sometimes the energy of the situation shifts without any apparent cause. As a Witch, I honour this process, but I also use the tools I have learned to ease the journey. There is no way of understanding some situations and the only way to move forward is to ask difficult questions and to cry it out.

I did not have my black mirror with me, so my friend and I joined hands and, by concentrating energy down through our fingertips into the space we held together, I created a black hole between our connected hands with visualisation. Although my friend is not a Witch, he was active in this work: together we held our fingers in the shape of a circle and willed the space black. Once the energy felt its way into the hole, I asked my friend to move to a standing stone not far away so that I could work on my own. While he guarded the circle and actively sent me energy, I focused the image of my dead friend into the space between my fingers. I was prepared to take as long as necessary to conjure up a mental image of her and in fact it did take some time. Every time thoughts got in the way, I refocused until I felt I had a connection with her. Then I asked some questions. I asked her to look in the mirror. Her part in the mystery was that she had not communicated something important to me. The result was that I felt unheard. There was no way to feel heard or complete without actually 'completing', myself.

For some time I felt nothing but the void. But after a while I felt the mirror working. I sent it as strongly as I could. In my mind's eye I imagined the mirror flying to the root of the problem. And there was a definite ending to the journey. The mirror flew to my dead friend and also to another who was an obstacle in the situation. I sent the mirror to its targets, saw it meet both of them and asked them to look at their actions.

I sent the mental mirror into the ground, bringing some light into the space of blackness between my hands, then, placing my hands on the ground and willing the mirror down, I wished my dead friend peace and asked for some healing myself. The circle was banished and as I walked away I felt some relief for the first time since she had died.

*What you start you must finish and in a consecrated circle this becomes all the more important. The mirror would do its work on a psychic level and the rest was up to me on a physical level. Later, whenever anyone referred to the situation, I asked the same difficult questions openly. I sent that energy forward as a healing. I am still asked about this situation and rather than keeping silent I speak about it openly as a way of honouring my magical path.*

# SACRED WELLS, HEALING WATERS

Our dependence on water is one of the strongest relationships we have with the Earth. Holy wells are some of the oldest sacred sites and formed a focus for ancient spiritual practices. Springs and underground reservoirs were home to animals and early dwellers who sought fresh drinking water, the most essential ingredient for survival, and the place where it came up from the ground frequently became the site of an early fertility cult. What began as a life-giving practicality became the core of spiritual contact with the Spirit of Nature.

Dowsers believe that most of the sacred sites of Britain were chosen as sites of spiritual practice because they were resting upon powerful energy spots in the Earth. The mechanism for finding such spots – dowsing – relies on a knowledge of water sources and the underground network of streams. Dowsers bear witness to the fact that most of the major sacred sites are situated on a spring and on ley lines, which depend on water for their efficacy.

In other words, if the sites are already powerful energy centres – working on the meridians on the Earth body, as acupuncture does on the human body – then sacred wells are situated on the points where contact with Earth energies is strongest.

Water devotion, the deification of the Earth as Mother and her waters as sacred, life-giving elixirs, gave rise to the many holy wells that cover the British Isles and other parts of Europe. Individual Goddesses governed certain springs and had particular powers ascribed to them. The Lady of the Lake, so famous in the Arthurian legend, was once the Water Goddess revered in nature. Lakes and rivers wove a map of the various local goddesses and gods associated with different sites. By the time Christianity reached Britain, water devotion had become so popular that many ruses had to be employed before the new religion could gain ascendancy over the beloved water spirits. Sacred sites that had been central to the old faith were often destroyed. In time, however, the Christians realised that there was a much more practical approach to the conversion of the Celts, as is evident from a letter written by Pope Gregory in 601 AD:

*When (by God's help) you come to our most reverend brother, Bishop Augustine, I want you to tell him how earnestly I have been pondering over the affairs of the English; I have come to the conclusion that the temples of the idols in England should not on any account be destroyed. Augustine must smash the idols, but the temples themselves should be sprinkled with holy waters and altars set up in them in which relics are to be enclosed. For we ought to take advantage of well-built temples by purifying them from devil-worship and dedicating them to the service of the true God.[1]*

*THE WELL OF THE HOLY CROSS, the Burren (previous page). Above, this well dedicated to Notre-Dame de 'Roncier in Brittany bears requests for favours similar to those that would have been left by our Pagan ancestors. Opposite, I walked through a sodden muddy field to find this forgotten Bridgit's Well in County Sligo. It was worth it – so beautiful.*

This tactic was broadly adopted as the way forward for Christianity, with the result that the old shrines and sacred places of the Pagans were renamed and 'owned' by the Christian God. Many people had no problem with making the same journey to the same special site and worshipping the new god as well as the old ones. For a time this was also true of former Celtic leaders, who formed political alliances with the Christians. The rural people were the last to convert, succumbing only when there was widespread persecution of Pagans.

# *S*t Bridgit's Well, Liscannor, the Burren

Bridgit's Well is perhaps the most extraordinary healing well that I have visited. It is a complex mix of Pagan and Christian references. The spring itself adjoins a burial ground on Church land: the well is a natural feature and the shrine that has developed around the healing waters contains a combination of plastic Catholic idols and Pagan healing ribbons. The result is a cacophony of religious imagery. The Virgin Mary rests beside carved pentagrams, symbols of Witches; healing prayers are said amidst candles, ribbons and plastic crucifixes. Bridgit is represented nun-like in a long black robe. Ribbons of various colours are tied to anything that will hold them.

This is a prime example of Celtic Paganism being subsumed by incoming Christianity. Bridgit is a Christianised form of Bride, the Irish Celtic Goddess of healing, poetry, inspiration, smithcraft and fire. The connection between Goddesses and their elements has changed over time: although Bride was a

Fire Goddess, she also had the triple forms of Maid, Mother and Crone. So great was her power that she could govern water sites as a representative of the three seasons of womanhood.

There is a lot of evidence to suggest that Bride was *the* main Celtic deity and thus her role as Water Goddess was a major one at any site. This well was certainly an important sacred site for Pagans. In time, here as elsewhere, Christians built their own architecture over the well and claimed the presiding deity as a Christian saint. Bride became Bridgit and in the process lost her sexuality. Nevertheless, the healing has continued until the present day and there is every indication that it will continue to flourish.

## BRIDGIT'S WELL MEDITATION

*Water is, of course, a great cleanser, both physically and psychically. This meditation, which I developed at Bridgit's Well, can be practised anywhere there is running water.*

*Listen to the water as it drips and rushes. Start to attune yourself to the sound of water on stone and the gentle gurgling. Breathe deeply into your belly, allowing the movement to be easy and relaxed – in and out, in and out.*

*Imagine a woman inside the well. She stands with arms outstretched, reaching forward to hold yours. She then guides your hands to the water. Imagine her placing them in the water. Wash your hands in the well. Allow the water to trickle between your fingers and feel the cool sensation of water on skin. As you do this, you are cleansing yourself psychically. The woman is a loving friend who has come to help you.*

*When you feel thoroughly cleansed, wash your face. Feel the refreshing waters on your face and smooth out any frown lines. Feel the texture of your face with the water on it. Let all doubts and worries peel away as the water soothes you. Let go. Breathe deeply. Breathe.*

**BRIDGIT'S WELL, LISCANNOR** – *a cacophony of Pagan and Christian imagery (left). Healing prayers are said amid candles, ribbons and plastic crucifixes. Above, beautiful worn stone steps lead down to an unnamed roadside well in County Clare. The local farmers tell me its waters make the best cup of tea in the Burren!*

***HEALING WATERS*** *Many wells have specific cures associated with them. In some cases the mineral content of the water has aided certain conditions. In others the simple fact of continually imbuing a sacred well with healing properties creates a place of immense healing capacities. Sacred water is believed to be 'charged' with the qualities of life force and purification. The charge comes from the act of devotion that takes place at the well. Teachers in the Craft speak of a place changing energy once it is loved, and this is the case with sacred water. Intention forms the charge. Once we link a place to our healing it becomes a centre for others to work with the same energy.*

*This is part of the lore of energetics in homeopathic doctrine. Experiments have shown that ions in water can change from positive to negative, with charging. This is done by sending energy through the fingertips into the water.*

*When healing, water can be used in whichever way suits the situation. If it is possible to immerse a wound in sacred water, then the wound will take energy from the well. Water also has the quality of attuning the chakras (see page 59): placing a drop on each chakra cleanses the energy centres of your body. If you drink healing water the effect will be both physical and energising. Many blessings that act as healings are given by water. The baptism of a baby is the descendant of the original water worship. To bless with the element of water is to sanctify your baby. It is to bless the one who has come from the birth waters and so we bless with the water of life.*

# Sacred Wells of the Burren

*Boireann* in Irish means a place of rocks. The place known as the Burren is an area of exposed limestone covering some hundred square miles in the north-west corner of County Clare. Great slabs of rock form rounded hills known locally as 'cow pats'. The rocks sit atop a wealth of natural springs which burst forth from the earth as wells.

*There are about forty holy wells scattered throughout the Burren. Not many are regularly visited anymore. But their locations and the cures they promise – for warts and feet, for delicate children, for toothache; stone chair for a backache cure, stone arch for headache and many, the great majority, for eyes – all attest to a faithful hope against elements and fate. Some have a very special mood about them; being nearby offers tranquillity which must always be part of any healing. And the water itself is a balm. In almost every instance, too, there is a family living nearby who have knowledge of the well and its powers, as if they were guardians of the place.*[2]

**THE TOOTHACHE WELLS** *on the haggard landscape of the Burren. It requires a combination of good weather and desire to find them.*

# St Colman's Well, Oughtmama, the Burren

This holy well is known to cure eye ailments and is still visited by locals on 5 November. The date may be connected with Samhain, the Celtic New Year, now normally celebrated on the night of 31 October, but which in ancient times might have moved a few days either way following the variation in the seasons. The well sits on the side of a mountain and is reached via a farmer's track.

It was information given to me by a local woman that enabled me to find it. Though Oughtmama is marked on the Ordnance Survey map, in reality these places are often hard to find, as there is no visible clue to where a spring emerges from the ground and no signposts. Mary, a local homeopath who is also the proprietor of the bed and breakfast I stayed in, had told me that the only indication of the position of the well was a leaning tree that stood out on the landscape. It was the tree that led me to the spring, for here on this rocky terrain, trees are few and far between.

The walk took me 40 minutes through green paths and barren rocks. It was a very quiet and satisfying walk, for just when I was beginning to wonder if I would be able to find the well, I cleared my mind and asked the landscape. Then I saw it. From where I was looking it was just a tree overhanging an overgrown stone wall. After feeling that I would never find the well it was a welcome relief to see the tree far up on the side of the hill. When I got nearer, I realised that without the spring, the tree would never have survived in this harsh environment.

It was my experience of this well and the spring within it that made me create a path into the underworld. In order to gain access to the well I had first to perceive it on the horizon, and then step down into the Earth on some lovely worn old stone stairs. Once I could see through the fuchsia and the lush greenery I could make out a spring coming up from the Earth within the well. This is a powerful metaphor, for it moves through Earth to Water. I often use the metaphors in nature to pathwork into the elements and to do deep personal work. In this case I had had to walk without too many clues to enable

me to find the place I was looking for. It appeared to be a well; but once I stepped down into the well I realised that it was a spring bubbling up from deep within the Earth. If I wanted to find a metaphorical link to the work I intended to do there, it would involve interpreting the ley of the land as a way of looking at my life at that point. To gain the wisdom I needed, I had first had to use a map and the help of others. Then I had to seek it on my own and, at the moment when I had almost lost the belief that I would find it, it presented itself to me.

We are often presented with a spiritual crisis when belief in the self and in a solution to a problem is challenged. Only by keeping the belief strong can we undertake the next part of the journey. What seems obvious is often not. The well appeared at first to be on the surface of the land, but then I realised I would have to travel deeper – literally into the depths of the Earth. I had to seek beyond the obvious. Only then would I find the source of what I was looking for. Water is the deep underlying current that yields many cures. I looked into a well that has a healing connected with vision – the eye cure – and I found my vision of life altered.

Surrounding the spring that came up from the ground were ribbons and religious artefacts. I washed in the spring and let go of all my immediate distractions. On that particular day I had come with a heavy heart and, feeling out of sorts, I let the journey take me to my healing. The washing was a psychic clearing, pouring all my sense of sadness into the well. But when I thought it was finished, I became aware of the metaphor of the spring within the well and I felt I had to go further. Meditating once again and this time feeling myself descend down the stairs into the earth chamber, I allowed my feelings to surface and draw up once more like the spring. Then I gazed into my reflection in the water for a long time. I felt the safety of the place and the security of life within the Earth.

**EYE-CURE WELLS** on the Burren (above and opposite, top). The natural spring opposite is shaped somewhat like an eye. Opposite, below, a horseshoe-shaped well on the Dingle Peninsula represents the birth canal or womb of the Earth Goddess.

There is a love that comes from places that have water meeting Earth. The journey to this well is worn by the centuries of visitors fuelled by a love of the Earth Mother. It is still visited, which means that it is active energetically. It is one of the hidden treasures of the Burren.

Once the deeper part of the vision had passed I stood up. I saw some tiny wild strawberries growing on the stones and ate one. Feasting at the well, I also drank the spring water. I had walked with intent to let the spirits of the place guide me. I had asked for a healing and for the ability to work with the deep well within myself. It is never predictable what will happen at a well, but if you are open to the sense of the sacred it can work its magic. I developed a meditation based on the site, to deal with the deeper, more personal aspects of the pilgrimage.

## PATHWORKING: THE SPRING WITHIN THE WELL

*When doing a pathworking, allow yourself to relax and breathe deeply into your belly. Really feel your body earthed and let the images flow through your mind. Be hazy and focus gently on the words. Sometimes it is nice to get a friend to read it to you. Or read it quietly with a soft mind.*

*Close your eyes and imagine you are at the base of a gentle hill. There is a track that is ancient yet grassed over, for it has not been walked recently. Along the path are some hedgerows filled with lush greenery. The sound of the birds singing is calming. You wander up the hill, for you have been told there is a well not far from here that has healing visions for the seeker. The track moves up along the ever-steepening hill, left and right it goes and you wander with it carefree. When you come to the top of the track you see that there is a higher mountain above and that the hill you came up was a foothill. The mountain is rounded and ash* white. At the base where the track goes all is lush, but as you go farther up, there is an ever-increasing moonscape of rocks. This is a wild and barren place, yet not without shelter. You keep to the base and look for the well, but it is not obvious.

The track stretches out along the base of the mountain and dotted across the landscape are rocks jutting out of the earth. You scan the horizon for some clue to where the well is located. It is not immediately apparent, so you take some deep breaths and ask for guidance. Then you look again and see something that makes you certain there is a well. On the barren landscape there is one striking tree that overhangs quite severely, as if the winds have blown it into place. You leave the track and make your way up the foothill. As you approach the tree you feel a sense of security, for there is life even in the most barren of landscapes.

*AT OUGHTMAMA* I stepped down into the well and saw little statues and offerings. Before I left, I drank some of the spring water and feasted on the tiny strawberries that grew nearby. It was all part of the healing I had asked for at the well.

Before you is a stone structure. At first you do not realise that there are some steps descending into a dark and overgrown place. Looking out over the landscape you descend into the earth and find yourself in a deep earthy place, overgrown with wild plants and flowers. There is a sense of peace within the well. Allow yourself to take in the deep peace of this warm harbour within the barren land and feel the love of the Earth for all that lives. Now you may wish to ask for some healing. As you breathe in tune with the ebb and flow of the environment, feel the warmth of the Earth surrounding you. Allow your deep inner feelings to surface.

As you become aware of this inner peace you realise that the foliage has become so overgrown that you cannot see the far side of this earth chamber. Pulling back some of the leaves you see many coloured ribbons and little images of a goddess. There are remnants of burned candles and many little votive offerings. Deep within the far corner you see a pool of water that comes up from the ground. Breathe freely and move towards the pool. As you look into the water you are filled with a deep sense of peace. This time it is the water that brings forth visions. Look into the water, seeing your reflection within. Watch as the images change and dance and presently you see the face of a woman who is the guardian of the place. She asks you what is the healing you want and what you would like to leave behind to the water. In silence you receive the healing of the water. Receive the healing

*energy into your body so that it lights you up in a golden shimmer. Fill every part of you with this healing light. When you have received your fill of the healing, thank the Lady of the Well and wash in the water. As you stand up you see that there are some wild strawberries growing along the sides of the well. Eat one and complete the healing exchange with the place.*

*The reflection of the woman in the water has faded and it is time to leave this place. You walk up the steps and make your way out of the well down the foothill. Your path down the hill is easy. The stones lie on wide flat expanses and there are cracks with flowers growing in abundance. You turn to look at the lone tree and see a white cow standing near the well. She is grazing contentedly and then looks up. You exchange a look of understanding and say farewell. Down the hill and back to the track you go. Follow the track back to where you started. Breathe and then gently come back to where you are.*

*When you feel ready, write down the healing you asked for and what it was that you left behind. Later it will be useful information to work into your awareness.*

The use of journeys as a metaphor for life is part of the work Witches do. We use pathworkings as a means to access the hidden psyche and to reach parts of the personality that need expression or release. The thoughts that these journeys bring to the surface are part of the release and inner guidance that you may do in the ongoing work of transformation. Ask yourself what you need in your healing and whether there is something you need to leave behind. Answer yourself as honestly as you can.

I find that the experience of walking the journey – whether literally on my way to a site or figuratively in the course of a meditation – with eyes wide open and receptive gives me insights into what I need to work on in my life. A shaman sees every journey as a metaphor for the inner journey. When I go to these places I ask as a Witch, for I know that the land heals and answers when we open ourselves to it. I left Oughtmama revived and with a sense that the landscape had worked its magic on me.

**ST COLMANS WELL, OUGHTMAMA** *I had begun to believe that I would never find it, but then I asked the landscape and the well revealed itself to me. The journey to the well was an integral part of the ritual that I performed there.*

# Innerwee's Backache Chair, the Burren

This is a place where a 'seat' has formed into the landscape and is reputed to cure backache. I feel it is a special site and in some ways a hidden place. In the summer it would be known only to those who peruse Ordnance Survey maps, but my friend Saira had seen it when it was less overgrown. Three women and Saira's daughter, Alafi, went on the way to the airport because Jan has had serious backache for many years. Even though she had a plane to catch we made our way through the undergrowth and she sat for a while on this chair. We all did.

**JAN AND ALAFI** *at Innerwee's backache chair. This stone seat has been reputed to cure backache for many hundreds of years, and it may be that the memories of other sufferers has passed down through the generations to provide healing energy to us.*

It is impossible to measure each person's response to this kind of sacred site. Jan's backache is an ongoing issue in her life and one that requires chiropractic treatment, but here in this quiet place she had a moment to reflect on what it was that had created the problem. Her

back is part of her journey and any relief en route is helpful. Perhaps Innerwee had backache herself and sat there in distant times. If two people in different times wish for relief for the same problem at the same place, it may be that this forms a psychic link that can shift the thought pattern that has caused it in the first place. This chair of stone holds memories of others who shared the trouble of backache and it provides a place of reflection.

Innerwee had connections to the famous sheela-na-gig over the entrance to the ancient church of Kilnaboy, just down the road. The exact nature of the connection is not clear, but we know that she was an abbess and the church would have been significant to the local community. If Innerwee was a strong female presence, perhaps the sheela-na-gig was part of her hidden teachings. Kilnaboy and Innerwee's chair are places ascribed to the female in mythology and may have formed part of Celtic Pagan oral history. Both have distinct female references which have not been suppressed, and this is unusual.

In *The Secret Places of the Burren,* John M. Feehan describes a strange experience he had in the ruins of Kilnaboy church:

*All this leads me to a beautiful summer's day many years ago when here in the remains of this holy place I sat by the wall of the graveyard with my beloved dog, Maxie, lying by my side … I closed my eyes, relaxed my body and drifted into a reverie. I slowly banished all thoughts and let impressions only have free rein. I was hoping that I would be able to roll back in time to the days when the chanting of psalms and the singing of hymns echoed through the dells of Kilnaboy. But it didn't work. It never does when one tries to force it. Instead there was one word, one impression jumping into my consciousness and it would not go away. It was the musical word 'Innerwee'. Was I going to love again? She slowly began to take shape in my mind, a thing of infinite beauty like a portrait that had been painted with loving care by an artist from heaven. Her presence became so powerful that I took fright. How crazy can one get, I thought. After all she was a nun and probably an abbess. I jumped up and made for the car, smartly followed by Maxie. As I drove down the road I turned my head the other way as I passed by Innerwee's chair, a rock just inside the wall reported to cure troubles of the heart.[3]*

**SHEELA-NA-GIG AT KILNABOY** *Sheela-na-gigs are outrageously and provocatively sexual, but, like other ancient Pagan symbols such as Green Men, they are often found in churches, carved by the followers of the old religion who worked there.*

If there is an inference that the cures for backache have a connection to 'troubles of the heart', then this place makes direct reference to it. As many healers look to the underlying cause of physical symptoms, mythology sometimes holds clues to these ancient connections. Certainly in homeopathic teachings and in Chinese medicine, the symptom is the tip of the iceberg. The whole person is taken into account before treatment is offered. In many old stories there are the traces of teachings about such connections. When a place is said to have certain healing effects and also has a reference to a state of mind there may be indications that the two are part of the holistic profile. Is then backache part of a profile of someone who has troubles of the heart?

Many homeopaths working from the school on the Burren have used sacred sites to do 'proving', by gathering the healing essence of a certain place. It is perhaps best described as shamanic homeopathy, but it shows that there are many ways of taking a cure. To increase the potency of any cure, take it at a sacred site and with intention. Ever important in any healing work is the conviction that what you are doing will work. Believe until you know it works. This is the power of intention and is the basis of any shamanic and magical work.

**POULNABRONE DOLMEN** (main picture). Right, visitors make small replica dolmens with the rocks of the Burren. Below, Toothache Well, a small basin in the rock, reminding me of an American Indian site.

## Toothache Wells

It requires a combination of good weather and the desire to find these special little wells. On the day I visited them I had both. The Burren in its glory changes with sunshine into something akin to a desert. I came upon these wells late in the day. They were marked on the map and I had information on the various wells of the Burren with me, but even so it took quite a search to find them, as there are many strange and irregular outcrops of rock on this flat plateau. Some of these strange formations are natural, but many appear to be the result of visitors piling up stones. I have seen this phenomenon nearby, at the famous Poulnabrone Dolmen, where visitors stack stones in a likeness of the dolmen, and in the instances of cairns dotted on the Neolithic landscape, but here it is less obvious. The stones seem to be a combination of the natural and the man-made. I once again had to become still and ask the landscape for clues. These came once I had opened up to them and I found myself in front of two truly primitive wells.

My first impression was that they were like an American Indian sacred site. The springs that came up to form small basins in the rock were surrounded by adornments and bits of pottery. In amongst the more obvious religious imagery were bits of plastic, chipped pots, string and coins. This was a place of healing, but severe

*COMMON PRACTICES AT*

*WELLS include throwing coins into a 'wishing well' and tying ribbons, known as clooties, around the well (see opposite). The coin tradition is one of the oldest of rituals and its origins are ascribed to Pagan times. Since it was once widely believed that water spirits or deities were connected to sacred wells, offerings were given to the water in the hope of pleasing them.*

*AT MADRON'S WELL the red AIDS ribbon now often appears among the more traditional clooties that have adorned the natural spring for centuries. Recently a local priest has got the Boy Scouts to 'clean them up'.*

*In Roman times it was the goddess Diana who could be contacted by throwing silver into the water. Diana ruled over the moon and tides, and when the Romans came to Britain she came with them. Her worship is remembered in the action of throwing coins (although they no longer have to be silver ones) into a special well or pool of water – in order to have a wish granted.*

healing. I could imagine the agony of toothache being equated with the hardship of the rocky landscape and felt grateful that I had no need for treatment of this sort. To get there in bad weather the sufferer would have had to endure a harsh journey, for it had no warmth. Here there was no call for Mary or the usual saints that are invoked for help in Ireland. Here there were rocks and sky and yourself for company. I could imagine it as a place of vision quests.

Perhaps it is harsh in order to meet the needs of the people who lived there, for in the Burren it is possible to see the cruel history of British rule during the famine. In an attempt to keep the peasants from revolting, the men were given a pittance in wages to build useless walls up the side of these ragged hills. The walls bear witness to lives wasted in a useless pastime. Walls that keep nothing in or out. Walls without purpose. Sometimes you can see trackways going nowhere, built for the same reason. They are a testimony to someone's skill and sadness.

*Many objects, including people, were once sacrificed to the water gods to gain favour. In London the River Thames has received many such sacrificial objects.*

*The clootie was originally a piece of rag tied to the branch of a tree near a well or at another sacred site and left to rot. If it was removed it could bring bad luck to the one who placed it there.*

*In his excellent book on Cornwall, Paul Broadhurst says:* **The old practice of leaving some article that has had intimate association with the diseased part is a ritual that is as old as man himself, and is to be found in every corner of the globe. The rationale for the practice is contained in elementary sympathetic magic, working on the principle that as the rag (Clootie) or article decays, so the diseased part heals naturally, in much the same way that wart-charmers sometimes rub a piece of meat against the wart and then bury it to decompose. The practice is still widespread in Ireland and is common amongst the Shintoist devotees in Japan.** [4]

*The tying of a rag or flag is evident in American Indian sweat ceremonies, Tibetan prayer flags, fetish religions of Africa and Indian shrines. I have seen many instances of it in many parts of the world.*

# Madron's Well, Cornwall

This beautiful well is set along a winding country path far enough from the road not to attract significant numbers of visitors, although it does have walkers passing by regularly. Ultimately this means you should be able to spend time there alone. The visitor's first glimpse is of an impressive array of multi-coloured clooties. The wells were here long before modern illnesses and have seen plagues come and go. For her devotees they remain a lasting symbol of a caring Goddess and represent hope in what can be an overwhelming world.

Further along the path is the Christianised version of the Pagan holy well. The natural spring has been contained by a beautiful ancient baptistery, an ideal place in which to perform a quiet meditation. As the baptistery is missing its roof, it is easy to hear the approach of another visitor. In one corner of the building is the sacred well, with constant running water surrounded by ferns and moss. The spout pours into a basin-like pool and this makes the water easily accessible. This is one of my favourite places to do healing.

*Madron* is Cornish for mother. One of the women I took there was very concerned about her mother and it seemed fitting to do a healing for her at 'Mother's Well'. Kate's mother had been working with ovarian cancer for two years. (I do not say 'battling with', as this suggests a war, rather than an understanding of oneself.) She had undergone several radiation treatments and had been very brave. Now she had lost her hair, lost weight and lost her spirit. An essential part of the equation was that Kate had been severely affected, having looked after her mother over the two years. The healing I wanted to do included both women.

## HEALING AT MADRON'S WELL

*I asked Kate to sit beside the well, listen to the steady flow of water and envision it as the flow of life force: gentle yet strong, always available. Mother Earth provides the waters of life to everyone and I asked Kate to get in touch with the vast healing Mother, to allow herself to flow with the water deep into her soul, to send life force to every corner of her body and mind – and to let herself go to the Mother.*

*Kate had had to be the family pillar of strength. Once she allowed a greater force – in this case the water – to be the strength, she could rest a little. The rest brought tears and I asked her to catch her tears in the chalice I had brought. I then asked her to use the same chalice to collect some water from the fresh-flowing spring to mix with her tears. I took her on a meditation in which the Mother's life-giving waters were composed of the tears of all the women in the world. Tears of happiness combined with those of sadness to become life-giving waters of birth and healing. Once Kate had moved through the tears, I asked her to place her sorrows psychically into the chalice by imagining her pain filling it. She then poured the contents of the chalice into the pool at the base of the flowing water and let go.*

**THE LATER BAPTISMAL** at Madron's Well is really a spring that wells up to create a small brook. Surrounding the brook are low-hanging branches where visitors leave offerings. Opposite, Saira bathing in the freezing cold White Cow springs, the Burren.

*She was concerned that this might affect the pool in a negative way, but I told her that the Mother is much bigger than us; so much bigger that Kate's pain and tears would only add to the life force that is the sacred water. I asked her to collect more fresh water in the chalice to imbue it with all the healing energy she could give. In doing so she imagined its healing force, never-ending and available to all.*

*Sending our thoughts of love and strength to her mother, we placed our fingertips into the water and willed healing into it. Then I poured half of the healing water into the pool for the Mother Goddess and put half in a glass bottle for Kate to take back to her own mother. This water could be used in many ways, but I suggested adding it to a vase and putting freshly cut flowers in the room where her mother slept. This is a subtle way of giving sacred water to someone, and it also means a lot to the person who places it there. Once the flowers die, the water can be poured into any flowing water or down the sink with the tap running.*

*THE WATER ELEMENT IN RITUAL AND MEDITATION* Water is tidal and is therefore linked strongly with the menstrual cycle. All issues of the Feminine can be addressed through water. It is birth, life, She, the emotions, love, healing, cleansing, clarifying and daring. Water flows, gurgles, bubbles, rushes and lies still. Changeable and fascinating, the waters of the world come in so many forms: rain, ocean, brook, lake, loch, stream, river, tears, rapids and sacred wells.

Whenever you wish to touch your spirituality, go to water, for there you will find the answer. Water can be used to wash, to purify a person and place, to soothe, to scry, to drink from a sacred well. In all these cases, the power of water works on a deep level. It can be used for any act of cleansing.

*To purify a place*

Take some of the well water in a glass or jug which will serve as a chalice. As 'well-water' is essentially sacred and imbued with the energy of the place it springs from, it can used to sprinkle the outline of a circle in which to meditate. It can also be used to create a protecting boundary round your own home. Starting at the front door, hold the chalice of water in front of the door and make a positive statement in your own words about safety – your safety. For example, 'I guard and protect this place, my home, that I may sleep soundly and safely in this place.'

Then, working always in a clockwise direction, sprinkle the perimeter of each room, going into every corner and continuing without stopping round the entire house or flat. At each window and door make another statement as you sprinkle.

Once you have arrived back at the front door you must seal the space. Create a mental picture of the flow of energy all the way round your house to the starting point and join it up in a continuous circle. Sealing is essential; otherwise you are leaving a hole in your consciousness about protection.

Then repeat the protective ritual using incense.

*If you want to do a blessing of the house with all the elements, then use incense (Air), a candle (Fire), a chalice (Water) and salt (Earth) in that order. Take each around the house into each room, always holding the element up to each window and door. Seal the openings psychically with intention. Witches use all the elements in ritual to address all aspects of protection. This is a way to fix intent but can be done with as much potency with one element. It is the intention and the energy given to that intention that creates the protection. Water on its own is particularly cleansing, as is incense. Fire can drive away negative influences with its purging quality. Earth is a great absorber of negative influences and has the quality of stability.*

*Once back at the start, go to whatever you feel is the centre of the house. For me this is my altar, which represents the psychic centre of my spiritual practice. In your own words, make a blessing on your house, always speaking positively. Say what the place is, rather than what it is not; for example, 'I centre myself at the centre of my house of love, compassion, abundance, health, contentment, safety. May this be a place of balance. May I be centred in this space.' Wiccans may say: 'Blessed be this time, and this place, and they who are with us.'*

### Spells for letting go

*In any instance of needing to let go of someone or something, water is a very gentle and complete element to help you.*

*First of all, think what use of water you wish to call to your aid.*

*Always there is the cleansing. Cleanse yourself with the well water. Feel it slip through your fingers and let go of all unwanted stresses and worries. Let the doubt slip away with the water. Imagine yourself clearing the path of all tired ways and thoughts. If you are able to do this at the wells you visit, it is a simple and powerful way to bring yourself into the now and let go of the past.*

*Water wakes you up. It is sensory and sensual, so feel the waters of any well you visit. Before you drink healing water, take a good look at it to see whether it is indeed still a natural spring or is a little stagnant. Some, like the white waters of Glastonbury, are used as drinking water, so you will know that they are OK. I usually go by sight, but even if you do not want to drink the waters, use the cleansing in an external way. At the very least dip your fingers in it.*

*The act of mentally cleansing at a sacred well actually clears your mind and spirit for other forms of water magic.*

### To let a lover go

*If you want aid in healing a wounded relationship, to complete with someone, to let go of a love relationship or to say goodbye to a loved one who has passed away – then use water.*

*Fill a bottle of sacred water and hold your hand over it, all the while sending your goodbyes into the water. Mentally say everything you wanted to say. If you are in a place where you can speak out loud, speak into the water. Allow yourself to cry and really let go of the past. Pour your feelings into the water. Sing or chant into the water. When you feel finished – and this must not be rushed – seal the bottle with a cork or bottle-top. Take the bottle to a place of running water and pour it away. Ask that the soothing spirit of the water receive your hurt and that in turn no hurt will be passed on.*

*Cleanse yourself ritually, concentrating on letting the relationship go. This is particularly effective if you run a bath and decorate it with rose petals and sweet-smelling oils. Lavender is a wonderful uplift and a few drops of the oil in the bath make it a fantastic ritual. Rose oil is especially good for women. Make your bath a haven in which to rest your tired body and focus your thoughts on letting go of that relationship as you wash from head to toe. Remember the song 'I'm gonna wash that man right out of my hair' – well, this is that song in action. Treat yourself as if you were the most precious object of beauty. Take care of yourself and wash yourself the way a mother washes her beloved child.*

*Once you feel complete, pull the plug out while you are still in the bath. Allow the water to pull away from your skin and let all traces of that past relationship slip gently away. Mentally complete with the relationship as the last drops of water disappear down the plughole. Then take your bottle of sacred water and bless your chakras by putting a drop of water on each and sealing the action you have just performed. You are purifying yourself, so it is important to bless each in turn. Genitals, belly, navel, centre of breasts, throat, eyebrow centre and crown. Bless each with a positive thought: 'I let go of X and heal.'*

*If possible, drink the sacred water as part of your completion ritual. I have drunk to endings as well as to beginnings. In a sense all endings are beginnings, and it is important to honour that rite of passage. Connecting with the birth waters is part of the nature of water. It flows and changes and when we let someone go we are acknowledging the change in our lives. Sometimes it is enough to meditate quietly on the letting go, then drink to the ending and the beginning.*

# S ACRED LANDSCAPE: THE GODDESS AT AVEBURY

The sacred landscape of Avebury in Wiltshire lies in an area full of ancient spirit centres. It is composed of many sites, within view of one another. At the centre is the vast stone circle, so large that an entire village sits within its boundaries. Within this massive circle lie the remains of two smaller stone circles. An enormous hand-made ditch and bank enclosure – a henge – surrounds the site. Two snaking processional routes wind across the rolling hills of the chalk downs to emerge in the central stone circle. Few of the stones that once stood in pairs lining these routes remain, but some of those that are missing have been replaced by concrete marker stones, so that it is still possible to appreciate the grandeur of these processional ways.

Two other sites (at least) are important in this sacred landscape – the West Kennet Long Barrow and Silbury Hill. The Long Barrow is unquestionably a burial chamber, but there has been much speculation as to the function of Silbury Hill. Seen as an overview, the landscape is a living Goddess with different sites deemed important at different seasonal festivals. The link between Avebury stone circle and the other sites has been described as an interplay of the seasons of life of the great Neolithic Earth Goddess and the fertility rituals that ensured prosperity to her

worshippers. Michael Dames, a leading authority on Avebury, considers the interaction of the ancestors with the land to be the foundation of a major centre of fertility worship. The existence of a sheela-na-gig carved into the side of the font of the church at Winterborne Monkton, just two miles north of Silbury Hill, is another pointer to the emphasis on female sexuality here.

As each season came into its own, local customs ritually referred to what was happening in nature. When the trees blossomed, fertility rites and the lust of new life was enacted. Sex was sacred. Even today, at the ancient festival of Beltane (1 May), a local woman may be chosen as the May Queen and adorned with flowers. This harks back to a time when the chosen one may have been a priestess who was perceived as the earthly representative of the Goddess for that year. In some of the fertility customs there is a suggestion that she may have mated with a chosen priest, or Stag King, and performed the Great Rite for the land. This sacred ritualised act of sex was performed on behalf of the community. There is a connection to the Great Rite in Wicca when a couple engages in sex in a ritual as a way of blessing or celebrating a season. The widespread belief that Witches indulge in orgies refers to these fertility rites or to a celebration of human sexuality and pleasure. In Wicca, to give pleasure is to honour the Goddess, though on a practical note the Great Rite is performed in private between consenting adults. It is a sacred act involving only two people.

*SILBURY HILL (previous page) is now generally considered to be a birthing site, representing the belly of the pregnant Earth Goddess. The ripe belly hill sits in the rolling downs, which in late summer are covered with wheat. The harvest is visible here. People live and work at Avebury (above and opposite), which lends the place a fascination that some other sites do not have.*

The seasonal rites followed both the agricultural cycle and the solar alignments. Distinct areas of ritual activity took place in different locations within the Avebury landscape. The coming of young sexual awareness, symbolically early spring, was the time for making ready the seedbeds.

Purification rites signified preparation of the land for the seed. The spilling of the blood of a cock into the freshly furrowed land (a farming ceremony that was still taking place in Norfolk up to the 20th century) was one way to express this theme.

If the land *was* the Goddess then the furrows were her sexual awakening and the fresh blood may have represented the blood of life or menstruation. Once the ground was prepared, the seeds were blessed and sown. New beginnings are echoed in the Wiccan and Celtic seasonal celebrations. Imbolg (2 February) is the time of purification and of new light. It is the hope of the year, before any ideas have been set or intentions focused. It is the time when the Earth Mother is prepared for the ploughing and seeding that will be done at Beltane. This was the time of instruction before the initiation of young adolescents. The coming of sexual awareness brought with it rites which acknowledged this threshold. Among Christians, even married couples were supposed to 'do it' only in order to reproduce. But much of Paganism was about looking to nature, and in nature, sex was everywhere.

# Silbury Hill

Silbury Hill is one of the first sites I visited in the context of a sacred journey. Many years ago, before the hill was fenced off, it was easy to climb to the top and see the interplay of Avebury henge, the stone circles, the long barrow and the avenue. Something that cannot be explained – and this is also true of Glastonbury Tor – is that the climb to the top is very easy. Of course not everyone will find it so, but time and again I have heard people say that they had been expecting a more tiresome journey. There is a kind of illusion about these hills. Once started, you are up there in a flash. It is part of the magic of sacred hills.

**VIEWED FROM ABOVE,** *Silbury Hill, rising steeply to a flat plateau, resembles a belly or an eye. With the moat filled with water, the appearance is one of a Neolithic squatting goddess – a goddess in the birthing position which was favoured for centuries .*

Silbury is one of the most impressive of all Neolithic sacred sites, for it was built using only antler scrapers and flint tools. Legend has it that a mighty king known as Sil, or Zel in the country pronunciation, was buried here on horseback and that the 'hill was raysed while a posset of milk was seething'. The mention of milk may be another reference to the ancient milk rites that recur in Celtic mythology. Though archaeology tells us that this is not a burial mound, the site remains an enigma, for neither is it just a hill. Producing a hill of this scale required both great spiritual belief and great organisation, which suggests that it had significance to the ancestors who constructed it.

Silbury Hill was created from the chalky downs it sits upon. It is composed of a series of concentric spirals that get smaller towards the top and from here there are grooves running down the sides of the hill. The grooves are linear and radiate out from the centre. This kind of symbolism is also seen at Dowth and Newgrange passage tombs (see page 148) and is suggested in the spiral art frequently found at Neolithic sacred sites. At the time the ancestors gazed upon this feat, it would have been outstanding in the surrounding landscape, for it was layered flint and chalk, making it startlingly white against the backdrop of the downs.

*BOTH THE BELLY AND THE EYE* are ancient symbols of the Goddess. Much has been written about the eye goddess of Mesopotamia and it is an image that recurs all over Europe. The eye motif has connections to the spiral of Celtic art.

*The womb and the eye of the world can also be seen as representing the navel of the world. In Neolithic art there are many examples of an image of part of the Goddess representing the whole. In some cases it is her head, but as ideas become more abstract we find the image of the eye, which is part of the head, signifying the whole Goddess.*

At the base of the hill is a deep moat that fills with water and is an essential part of the site. Seen from above, the hill can appear to be several sacred images, including that of a squatting Goddess. The paradox is that even if the people who created it did not have the means to see it from the air, they seem to have had the perspective to imagine how it would look. There is much speculation as to whether this means that they had some form of flight. Some of the more vivid explanations are that visitors from another planet designed the site, or that shamans were able to 'fly' with the use of herbal potions so that they could envision an overview of the area. Many other sites throughout the world pose the same questions, the most famous being the mysterious Nazca lines in Peru, which are so large that it is possible to see the whole image only from the air. Silbury, Michael Dames concludes, is a universal image of the Goddess. Whether the rounded hill is representative of her full belly, her pregnant belly or indeed the all-seeing eye, the land is the Goddess.

Nearby is the Swallowhead Spring, the source of the River Kennet. The rise of a spring from the Earth was a wondrous thing to the ancients, often seen as the source of life-giving waters. The distinct female imagery of the waters issuing forth from a dark cave, as it does at Silbury, made a natural Goddess site. As the stream runs over the chalky bed it turns a creamy white like the waters of ovulation – a phenomenon also seen in Glastonbury, with its White and Red Springs. White springs suggest fertility and these waters come out of the sacred birthing chamber cave, perhaps symbolising the place from which divine life emanated. Surface water is rare in chalk country, so this would have been an anomaly at a time when people were looking to the land for inspiration.

**THE WHEEL OF THE YEAR** turned and the people moved with their symbols of what was true of the season. Thus spring, when flowers emerged after the dead of winter, was seen as a time to enact seed or ploughing rites. Harvest was a time to reap or cut. The recurrence of enigmatic crop circles at Avebury does not stop local farmers from harvesting.

*MICHAEL DAMES speaks of Silbury Hill as the pregnant belly of the Goddess, and the cave and its spring as her pussy or cunt. The word cunt itself may have origins in the pre-Christian name of the Goddess associated with the Swallowhead Spring. The word that we now hold as the most vulgar in the English language was once a revered and sacred name. The 16th-century topographer John Leland emphasised the connection:* **'Kennet riseth at Selbiri hille bottom'.**[1] *More recent research tells us that as late as 1740 the River Kennet was known in local dialect as the Cunnit. It has been suggested that this is the oldest, most sacred reference to the female genitals, the power of life and the Goddess.*

*In support of his theory, Michael Dames points out that the oldest relatives of Neolithic European peoples are the Basques. They are pre-Celtic in origin and their word is* Kuna. *Some older forms of this word are:* Cunnus – *Latin English;* Cunte – *Middle English;* Kunta – *Old Norse;* Kunte – *Old Frisian;* Kuna – *Basque. Words that derive from these sources include:* Cunabula – *cradle, earliest abode, the place where everything is nurtured in its beginnings;* Cunina – *the Roman goddess who protects children in the cradle;* Cunne – *to inquire into, explore, investigate, to have experience of, to prove, to test, to taste;* Cunning – *knowing, possessing a practical knowledge or skill, able, expert, dextrous, clever, possessing a magical skill or knowledge.*

*WILLIAM STUKELEY, an 18th-century antiquarian, mapped out much of the site of Avebury. This is his overview of the snaking processional paths leading to the circle which represents the central sexual metaphor.*

*All the words that were used originally to describe wisdom and magical knowledge have somehow been turned on their heads. Where once they were sacred, now they are pejorative, obscene or dangerous. The Cunning Woman or local wise woman possessed knowledge of herbal cures and was consulted for her wisdom. But cunt has become derogatory rather than life-affirming. This is no accident. As we have seen throughout history, the Goddess-loving people who followed the old religion clearly had a reverence for the sanctity of woman. It is the patriarchal religions that have made these words obscene.*

*The history of Christianity is one of men talking a lot about men's achievements. The fact that the God is male has a definite impact on many people's view of 'the spiritual'. When I use the word Goddess, I am redefining our concept of the spiritual world. It includes women. There is very little in our language that speaks to women. We have few positive words for our bodies. If we are sexual, it is unseemly or sluttish (a very old word indeed). When I sat down to write this book I thought for some time about what word to use to describe the springs that are connected to Goddess sites. Vagina originally meant 'sheath', which in essence describes the female sex in relation to the male. I prefer pussy because it is slang and friendly, but other women find it vulgar. So what are we to do with our kunas, cunnits, cuntes, kuntas and River Kennets? Perhaps it is time to invent new names or to reinvent the cunt.*

# Avebury: The Processional Routes

Whereas Silbury Hill is associated with birth and pregnancy, the Avebury circle is the site in which the symbolism of conception or the heterosexual sex act is manifested. In ancient Goddess culture the mating of the Goddess with her consort is more than sex. It is the honouring of the Goddess and God, whose union brings forth a good harvest and fruits of the land. The ancestors who built these great sacred sites would not have considered the act of union to be the 'event' by which all others were measured. Many thresholds of human experience were as important. Different sites dictate different rituals, as each tomb, circle and avenue is suggestive of a special use of sacred space. At Avebury, the processional routes lead into the circle and may have been used to bring people

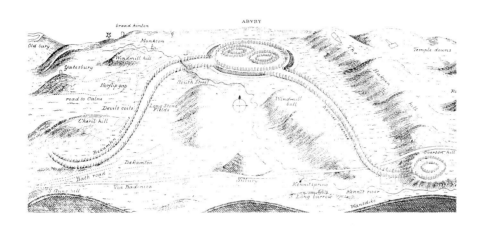

together to enact a central ritual for the community. The shape suggests the symbolism of a womb or container where creation takes place.

One of the ancient representations of lust and creation was the snake. At Avebury, as at Callanish, avenues of snaking stone alignments form a route to the central circle. These stones may function in several ways. They may indicate the direction in which an astronomical event would happen, or they may be processional routes, guiding the seeker into the central mystery. The walk to the site where a ritual was to take place was part of the rite. In some cases, the direction in which people entered or departed from the sacred place would have been symbolic. There are many interpretations; simply ascribing one avenue to young men and one to young women is only one possibility. There might have been times when young boys and girls made the entrance separately from their elders; alternatively men might have been separated from women, or the priestesses from the other participants in a rite. There are many possiblities where sexual, gender or age distinctions are represented in ritual.

At Avebury, the winding avenues, one arriving from the south and one from the west, once had 100 pairs of stones each. Dowsers interested in the construction of these avenues have concluded that they follow the main ley lines that intersect in the sacred landscape of Avebury.

*THE PROCESSIONAL PATHS AT CALLANISH*, like those at Avebury, lead into the central circle where the mystery is enacted.

Stone alignments and avenues have their meeting point in stone circles. Two winding avenues that intersect at the point of a circle are a powerful representation of a snake-like body meeting a circular mouth. This is the ancient image of the sacred snake of regeneration. It was once believed that the adder could reproduce by fertilising its mouth with its tail. The image of the snake eating itself or making love to itself is one of a hermaphrodite, for the tail is the cock and the mouth the pussy. One of the variants on stone rows and avenues is the cursus – a ditch and bank-like structure creating a path that can be followed across the landscape. The cursus near Stonehenge is 9000 feet long and contemporary to the building of Silbury Hill, which makes it about 4000 years old. Michael Dames talks of the sexual imagery of the cursus, suggesting that it maps the springtime journey between death and renewal. He also mentions the snakes' ritual disappearance into holes in the ground in the late autumn and their return at the beginning of spring – imagery whose sexual connotations are obvious.

The walk along an avenue of stones conjures up images of a sacred journey. This may have been the precise intention behind the building of avenues. It was a path of meditation en route to a ceremony, representing the path of life. To traverse a path with intention and to imagine it as a journey of the spirit into the mouth of the snake is activating a strong spiritual event. It is the journey

down the birth canal into consciousness. The processional avenue evokes a reaction similar to what one experiences when looking at great art – it may not be fully understood, but it can be felt deeply. If these routes were used as meditation paths or symbolically represented the path, then stepping on to that path with intention was a powerful ritual device and may have been used to raise consciousness in a ceremony.

At the centre of the Avebury Henge are two longstones that tower above the others. They are now known as Adam and Eve, clearly a Christian rendition of an earlier concept of male and female. The stones within the circle also suggest a pairing of two different shapes. The diamond-shaped 'female' stones stand by the tall thin 'phallic in intent' stones. This interpretation may be just a way of explaining describable difference, but the long history of the importance of fertility rites suggests that the male/female image was intentional.

This circular shape where the snakes meet is a place where all is possible. Here the central mystery was enacted and though we do not know what it was, the fact that stones are called Adam and Eve suggests sacred sex rites. Much later the May rites were enacted in one of the smaller circles with the erection of a maypole (phallus) and dancing around it. This too is a reference to fertility rites. Sex and fertility are also suggested by the much later folklore associated with certain stones and by the attempts of 17th-century Christian do-gooders to destroy them by lighting fires at the base of the stones and, once hot, pouring cold water on them to crack them. There are records of other, earlier attempts to destroy the stones and one is particularly curious. Under one of the great stones was found the skeleton of a barber surgeon, distinguished by his tools – scissors and an iron lancet or probe. It appears that he was employed to cut the devil out from the stones or perhaps to use his skills to change the Pagan site surgically. The stone fell on him and he lay buried beneath it until the 20th century.

*CROSS-DRESSING is significant in British folklore and in mummers' plays at Midwinter. The oldest customs include an important character known as Dirty Bet who is a man wearing women's clothes and is considered to bring good luck. Bringing in the new year and new energy was suggested by this lucky character who was 'between the worlds', crossing the threshold. The history of the Morris dance yields many earlier images of the woman/man character, in which again the transgender characters were usually males dressing as females. The faeries also occupy this special place: they are described as being androgynous, as are the shape shifters who are part human and part animal or fish. They cannot fit within the confines of the separate male and female identity.*

# The Devil Stone

The 'Devil Stone' has been renamed by Christians, for the concept of the Devil is not a Pagan one. In many records of Witch trials the word devil was substituted for the name of the local deity. But in fact the Horned God of the Pagans – known also as the Green Man, Robin of the Woods, Old Horny and many other names – was the god of the woods and the bringer of strong male sexuality. Thus the Devil Stone was, in essence, the stone of fertility. It was the seat of carnal knowledge, believed to impart sexual fecundity to those who sat in it. Up until at least the 1930s it was visited on May Eve by young girls who were to take part in May Day festivities the next day.

These girls were following the earlier ritual enactment of the path of the Maid in the Triple Form Goddess who is 'one unto herself'. There are stories of women and men going into the woods to collect greenery – whoever they

*GIRLS STILL SIT ON THE DEVIL STONE on May Eve, as they have done for hundreds of years. This is reputed to send sexual vibrations that encourage fertility in both women and men. I sat on it too.*

chose to meet in the dark was the lover for that night. This custom has variations such as meeting a future husband, meeting a lover while in a marriage, conceiving a May child and meeting a faery and making love. Of all the Celtic festivals Beltane is the most distinctly sexual, a traditional time of choosing or wishing for a lover.

Michael Dames suggests that the placing of the Devil Stone at the entrance of the mouth to the 'female' avenue makes it an image of the deity as transcending gender. The term Goddess was understood to comprise both the masculine and the feminine experience. The snake is a symbol of homo- as well as hetero-erotic experience, for it was believed by the ancients to eat itself and to have both male and female genitalia. Thus Avebury is the expression of all sexuality in a single image.

The placing of stones was always carefully calculated and stones of ritual significance were set in special positions. Thus, Dames says, siting the Devil Stone at the point where the phallic tail would meet the pussy-like opening of an inner circle was crucial to its significance. It was a stone that had within it the seeds of all sexual expressions: same sex and opposite sex.

There is a relationship between the symbolism of the stones and the customs performed there. Like the imagery of the hermaphrodite seen in the stones of Avebury, in the Celtic seasonal festivals there are characters that transcend gender. It is hard to assess whether the ancient attitude to hermaphrodite characters was positive because bisexual relations were more acceptable then. I choose not to call it an issue of homosexual or heterosexual imagery. I think it is more likely that these characters were simply considered outside the rules set by the later Christian culture. They were jesters invited into homes on Midwinter to 'bring in the new', which meant bringing in the light that would begin to return after the Winter Solstice. They had potency in their ribald and raucous natures and they pushed the limits to induce states of wild celebration and revelry. Bawdiness was part of the turning of the year.

# $\mathcal{T}$HE ISLE OF AVALON

If ley lines can be seen as the meridians of the body and holy wells occur at powerful energy centres, then Glastonbury is the heart chakra of Britain and lies at the centre of some of the most interesting Goddess lore. It is between the worlds, a place of magic where anything is possible. It is known in legend as Avalon, the land of apples that became the resting-place of Arthur.

Legend tells of Arthur's last fateful battle with his stepson Mordred, after which the king is taken by boat to Avalon. The versions of the story are so various and their mixing of myth, legend and history so complex that there is a mystery surrounding Arthur's end. Geoffrey of Monmouth, the 12th-century chronicler to whom we owe much of our knowledge of Arthurian legend, tells the story thus:

*The island of apples which men call 'the Fortunate Isle' gets its name... because it produces all things of itself; the fields there have no need of the ploughs of the farmers... Of its own accord it produces grain and grapes, and apple trees...*

*There, nine sisters rule by a pleasing set of laws those who come to them from our country. She who is first of them is more skilled in the healing art, and excels her sisters in the beauty of her person. Morgen is her name, and she has learned what useful*

*IN PAGAN TRADITION,* Arthur or Artos is the classic fallen or sacrificial king, as is Jesus. The concept of the sacrificial king appears in the seasonal festival of Lammas (1 August), when he is often represented as a god of the land or Corn King, the image of the ripe corn which must be cut at the height of its potency in order to give rise to a bountiful harvest. To allow the corn to become over-ripe or rotten will result in famine. The myth points to a stage in the god's maturity when he knows it is time to step down.

Like many sacrificial gods, Arthur is wedded to a Flower Goddess and his marriage is a sacred mating rite for the land. It is significant that his wife Guinevere is unable to produce a child. Symbolically, this means that the mating of the land does not take place and the Goddess is not tied to the sacrificial king. Guinevere's love of Lancelot is approved in some measure by Arthur, for in the tradition of the Flower Goddess she is not tied to one man but may chose whoever she wants. Her love is part of the health of the land.

*properties all the herbs contain, so that she can cure sick bodies. She also knows an art by which to change her shape, and to cleave the air on new wings like Daedalus; when she wishes she is at Brest, Chartres, or Pavia, and when she wills, she slips down from the air onto your shores…*

*Thither after the battle of Camlan we took the wounded Arthur, guided by Barinthus to whom the waters and the stars of heaven were well known. With him steering the ship we arrived there with the prince, and Morgen received us with fitting honour, and in her chamber she placed the king on a golden bed and with her own hand she uncovered his honourable wound and gazed at it a long time. At length she said that health could be restored to him if he stayed with her for a long time and made use of her healing art.[1]*

The legend leaves an open ending as to whether Arthur's wounds are fatal, but there is a suggestion that if he is to be healed, he cannot return to continue his kingship. He resides somewhere in the Mists of Avalon.

In Glastonbury the old and the new religions converge because the town is both Pagan Avalon and the supposed site of the first Christian church in England, established by Jesus's uncle, Joseph of Arimathea. The story goes that Joseph thrust his staff into the ground, and a thorn tree sprung up whose descendant still grows in Glastonbury. Joseph was a sea trader said to trade in tin from the Cornish tin mines. Rumours abound that Jesus accompanied his uncle on a trip to Britain and Glastonbury.

Arthur is directly linked to the old religion by his choice of the Druid Merlin as his chief advisor. Arthur outwardly accepts Christianity and unites the warring tribes under the new Sun God, but at the end of his life he returns to his allegiances to the old ways and to the Goddess. Although

in Christian tradition the Holy Grail is the cup from the Last Supper, the quest for the Grail is a reference to the ancient religion and to the cauldron of knowledge, otherwise known as the cup of the wine of life, or the Goddess. The cup is the metaphor for the ultimate knowledge of life, the womb from which all life emerges.

Glastonbury is described as an island because the waters of the River Severn have often flooded it. This marshy area was famous in ancient times, for when the Romans invaded it became a hideaway for the Celts, who knew their way through the swampy lowlands. Roman descriptions referred to a people who were painted blue, naked and adept at mercenary warfare. They were marsh dwellers and the Romans in their heavy armour were unable to cross this sacred landscape. The areas of Britain that have retained their Celtic flavour tend to be those that were hard to access: from Glastonbury the Celtic lands spread west down to Cornwall and into Wales, which had impenetrable mountains.

**IT IS SAID THAT AVALON** *(previous page) rose out of the mists and could be reached only by boat. This image of the morning of the total eclipse shows the mysterious isle surrounded by mist. Above, cleansing crystals in the Red Spring of Chalice Well.*

On the outskirts of the town are to be found the remains of Bride's Mound, named after the Celtic Triple Form Goddess. Here it is said that there once lived a community of women, variously described as priestesses or nuns, but always as healers. This forgotten site is part of the interlinked sacred landscape of Avalon. In Glastonbury we find the interplay of several Celtic Goddesses who are remembered in Arthurian mythology. In addition to Bride, Guinevere, Morgan-le-Fay and the Lady of the Lake all reside here. The tomb of Arthur and Guinevere is said to lie in the grounds of the Abbey.

The Tor, which dominates the skyline from whichever direction you enter Glastonbury, is the hill on which the tower – the only remaining part of a church that was destroyed in the Reformation – rests. Tor is the old Norman word for tower, but the hill itself has come to be known by the same name.

**THE HIGH IRON CONTENT** *of the Red Spring or Blood Spring gives the water its colour, and makes it taste less pleasant than the waters of the White Spring, which the locals drink in place of tap water. Opposite, the Maid – a Pagan woman at the Goddess Conference.*

Legend also mentions a maze on the Tor, which was once walked by priestesses. There is some evidence to show that the maze referred to is the snaking ley line which joins Glastonbury to Avebury and Stonehenge. Walking the ley lines and enacting rites along them may have been part of Goddess worship in Avalon. This famous ley line is known as the Michael line because of the number of churches on its path that are dedicated to St Michael. The church on the Tor was one of these.

The Michael ley line is one of the largest and most important in Britain. It passes through the Tor, connects it to the healing White and Red springs of Chalice Well and continues onward all the way to Land's End in Cornwall. Like other ley lines it is believed to be a natural pathway for energy or psychic information, known to the ancestors of Pagan religions, and is composed of a number of significant sacred points that join to form a line. With the influx of Christianity, it is interesting to note the number of Goddess sites that were renamed as Mary Chapels. It is believed priestesses or nuns who lived their life in the service of the Christian remnant of the Goddess – Mary, the Great Mother Goddess – originally dedicated their places of worship to her. However, these small chapels housing communities of women were subsequently reduced in stature with the predominance of male-focused Michael churches. What is of interest to feminists is that images of St Michael killing the dragon portray the phallic spear or lance being used to crush the Goddess, symbolised by the dragon. The dragon or snake resides at the heart of Glastonbury lore, for she entwines herself around most of the sacred sites here. As we have seen, dragons, snakes and worms are a reference to Goddess

spirituality and may have been linked to the snaking movement of the sacred ley lines. At the time of Arthur's conception, the prophecy received by Merlin referred to two dragons fighting underground – the red and the white dragon. As ley lines are often connected to underground springs it is also of interest that Avalon has two springs, a Red and a White.

Within the grounds of Glastonbury Abbey is to be found one of the remains of the old religion, the Egg Stone or Omphalos. There are no records of what this stone was, but clearly it is Pagan. It is shaped like an egg with a small basin in its surface, and it is likely that the basin would have been used for offerings, especially in the form of liquids. These may have been the milk offerings of faery tradition, or sacred waters from the wells being offered to the Goddess. In many ways liquid in ritual is a reference to the blood of life or the waters of birth.

The King of the Faeries, Gwynn-ap-Nudd, lives in the Tor itself and symbolises a potent male shadow figure. He guards the entrance to the Underworld and runs amok with creatures of the night in a midnight ride with obvious sexual connotations. This is the Wild Hunt of Celtic fame when an unsuspecting mortal encounters the Faery King and is carried along with the rush of the hunt. Those who are in Gwynn's path are swept along to the Underworld. In fact this is more a description of the loss of control one experiences when stepping into the shadow realms. Whereas the faeries inhabit the shadows as a place of safety, the moral of many stories is not to let loose the wild animal side of the

**CELEBRATIONS AND MEDITATIONS —**

*Chalice Well adorned with flowers for the total eclipse, 11 August 1999. Opposite, the night before the eclipse I conducted a rite for a group of women at Chalice Well.*

personality that was so out of keeping with the more restrained Christian ethos. Much of Paganism is about embracing both the dark and the light rather than denying the dark. In the dark one turns to face the self and there may be a discomfort in discovering the secret places of the soul. The path is less illuminated, but not more dangerous if one uses instinct and not the rational all-seeing clarity of light.

In Arthurian legend faery women offer help and teachings. Morgan is a healer as well as a trickster. But as with the mermaid tales, the woman who is free of society's limitations and is seen as a sexual being is regarded as a threat, if not as a Witch. In this case Morgan uses enchantment to seduce her half-brother, Arthur. It is sex and enchantment, embodying the young Arthur's loss of control, that assure his downfall, for this act of incestuous passion conceives Arthur's mortal enemy, Mordred. This account of the earliest of taboos was a Christian rendition of the punishment of sexual abandon.

Like Eve, Morgan is connected to apples, for Avalon is the Isle of Apples. Morgan of the Apple becomes the image of the Wicked Queen who offers the apple to Snow White. The poisoned apple is the apple of knowledge. Witches nowadays refer to the apple when in ritual they cut apples in half, sideways, so that a five-pointed star is revealed. This is the pentagram of the five elements in Wicca and is used in ritual at times of imparting knowledge. To cut the apple and offer it to another is a symbolic act on a grand scale, for it is imbued with thousands of years of Christian dogma: Eve offered the apple and Adam took it. In Glastonbury the orchards still exist and it is fine apple country. Perhaps the knowledge is the sense of ancient wisdom that comes from the sacredness of the place, and indeed the Tor and Chalice Well.

# Chalice Well

In traditional language visiting the Tor and the two wells of Glastonbury is seen as the Sacred Marriage of the phallic Tor and the Earth Goddess who sends forth life from the spring. Chalice Well is set within beautiful gardens and for a nominal entrance fee people are encouraged to spend time wandering and meditating.

It is a place of cleansing, of spiritual love and of emotion, and has long been a site of female ritual and healing. The high iron content in the water turns it a rusty red colour, lending the well its alternative names of the Red Spring, Blood Spring or Menstrual Spring. It is a traditional place to take the waters, which are said to be curative, even miraculous, and there are glasses placed so that visitors may drink.

The well works its magic on all who sit for a while. Walking up through the gardens one passes between two famous old yew trees, known as Gog and Magog, which form the traditional entrance way for the pilgrimage to the well. From the base of the incline the waters come down into a pool that has been made to look like a pussy. As you wander up from this first site, there are two areas which lend themselves to specific meditations. One is dark and the water flows into a stream where people often actually step into the water or wash crystals. Further up there is a more open, light area where the waters are drunk. The final stage of the journey is Chalice Well itself, a beautiful well whose oak lid has a wrought iron symbol of the vesica pisces, two circles that join to form a vesica or pussy. The tradition of the well as a site dedicated to the Goddess is upheld very literally here and the imagery is unmistakable.

*PARTICIPANTS at the conference used bones to make an effigy of a woman, carting her to the springs and up the Tor as an emblem of the end of the century. Opposite, 11.11 a.m., the zenith of the eclipse. Women light candles they 'charged' in the rite the night before.*

Just over the wall from Chalice Well is her sister spring, the White Spring, whose water is regarded as delicious by the locals and drunk instead of tap water. (Because of the high iron content, the waters of the Red Spring taste less pleasant.) The White Spring is considered to be a fertility site, with the white waters representing the waters of ovulation. It is surrounded by clooties that are tied to the branches overhanging the spring. Though this is less a site of meditation than Chalice Well, the stream runs through a beautiful well house where an altar is decorated with flowers for the seasonal festivals. Here, one can take a drink from a chalice that resembles the image of the Holy Grail of Arthurian mythology and is a direct Pagan tribute to the Goddess.

## GLASTONBURY TODAY: THE GODDESS CONFERENCE

*The present-day town continues with tradition and is home to many healers and alternative medical and spiritual practices. In August 1999 I attended the fourth annual Goddess Conference here. The organisers, Tyna Redpath and Kathy Jones, originally intended the conference to be a one-off event, but the interest was so great that they now feel it will be a regular occurrence. I was there for two reasons: to run a workshop on the Celtic Goddess, Bride, and to witness from the Tor the total solar eclipse of 11 August .*

*Glastonbury is definitely a special place. Women and men from all walks of life converged here for the conference. The theme this year was the Crone of the Old Millennium, so many older women were honoured in a ceremony called the Crowning of the Crones.*

*The night before the eclipse I had been asked to do a rite at Chalice Well for about 30 women who were on a tour from the United States. The three parts to the garden made it possible to layer the ritual into three events. The first was to invite everyone to arrive through the arch of the ancient yew trees, Gog and Magog, and be censed (cleansed by having the smoke of burning incense wafted around them). Each individual was then blessed with water to the forehead and invited to wash away any blocks, obstacles or disturbing thoughts. As they washed and many cried we asked them to tie a clootie on to the tree above the well. Once the tree was adorned with ribbons of letting go we walked in silent procession to the central well. Because it was the Year of the Crone, I asked the Crones present to stand around the well at the heart of the spring.*

*Earlier in the day we had asked the Crones to send a healing that could only come from a Crone to a Crone, the energy of letting go. This was because we had heard that Doreen Valiente, who had been meant to be at the centre of the Goddess Conference, had been taken seriously ill. As a result there had been one ominously empty chair at the conference that spoke to me of a dear friend letting go. We asked the Crones in our ceremony to send some energy to Doreen and spoke some words about the waters of the Goddess uniting all women: the waters of birth and the waters of life. We are made of water and united in it. Tears of joy and tears of sorrow come together to create one vast ocean of love. We sent that energy to our friend.*

*Each Crone then spoke some words about love, which was a wonderful experience for those of us non-Crones listening. The younger women heard the wisdom of the Elders around the well.*

*Once we had charged the water with healing we each carved a symbol into a floating candle that we then lit from the one candle at the centre for Doreen. We walked in silence to the next part of the well, where we took a drink of the healing waters. The final procession was to a place where we floated all the lit candles together. It was truly a beautiful sight.*

*It was important for me to take into account the eclipse that would occur the following morning and so each of us took away a candle. At the time of the eclipse I was back up on the Tor, but not all the women who had taken part in the rite the night before were together. Knowing that this would be the case, I had wanted to have a visible symbol of our connectedness, so that at the zenith of the eclipse we all lit our candles in the darkness and connected psychically.*

*As the moon obscured the sun and the shadow was cast, a group of us from the rite, who happened to be in the same place at the top of the Tor, stood in a circle and lit our candles. As we did so we chanted an om sound. The result was that at the point of total eclipse the whole Tor seemed to come alive and chant with us. Though there were hundreds of people there, the sound created by this small group created an opportunity for all who witnessed the eclipse to join energies. It is a moment that I shall always remember.*

**CROWDS GATHER ON THE TOR** as the sky darkens on the morning of the eclipse. We drank toasts to the Gods and Goddesses of Avalon. Inset, Randel Mars – wild Texan High Priestess – drinks from a chalice.

# PASSAGE TOMBS AND CHAMBER TOMBS

The burial sites of the ancestors of Britain are variously described as passage tombs, dolmens, cairns, long barrows and chambered tombs. These terms refer to the ancient burial mounds that were built from megaliths, enormous stones placed together without mortar to form a chamber in which bones were placed. The most interesting feature of the chambered tombs is that they were aligned with solar events such as the Midsummer sunrise, the equinoxes and Midwinter, and positioned so that a ray of light from the sun entered the inner sanctum of the tomb. Tombs with a passage to a central chamber are believed to be the living image of the Earth Goddess, who received the life force of the sun to ignite the chamber of the ancestors. The method of burial was similar to that used by American Indians, who placed bodies on wooden platforms so that the bones were picked clean by birds, and then laid certain bones in the tombs. In Tibet, the dead are taken to rocks, chopped up and the pieces left for the eagles to eat. Bones such as the thigh-bones and the skull are then taken and made into ritual objects.

It is not clear who was placed in the tombs in Neolithic times, but it is surmised that they were people who were regarded as important, possibly priestesses or priests or others who had some standing in the community. At the Tomb of the Eagles in

Orkney, only 342 individuals were buried in the course of a millennium and a half, and figures at West Kennet make it clear that by no means everyone was buried in a passage tomb or long barrow.

It is a feature of the chambered tombs that they appear to be in the likeness of a Mother Goddess and that the light that enters does so through the birth canal. Seen from above the interior is like a woman with her head as one chamber and her arms or breasts as side chambers. The leg chambers have between them the passage that is the entrance to the tomb. The image is of the tomb as a living Goddess, from whom all life emanated and to whom all life returned. The head was often the point that the ray of light would hit and in certain tombs such as Newgrange the triple spiral is lit up at the time of the Winter Solstice. In the later Celtic culture there was a cult of the head, which may have been seen as the house of the soul or the seat of the personality. The chamber at the head of the chambered tombs would have been particularly important. So too was the central chamber, the centre or womb of the Goddess. Most of the bones found in such tombs are not in the central chamber, but in the adjoining side chambers. Like the centre of a stone circle, the centre of the tomb was the place where the transformation took place in any ritual.

*CREVYKEEL COURT TOMB, a magnificent example of a court-style tomb in County Sligo, Eire. Opposite, the Tomb of the Eagles in the delightful wild setting of South Ronaldsay.*

It is also believed that the tombs were used ritually for many other purposes in addition to burial. Fires were lit and gifts offered, which may indicate the concept of an afterlife. In some tombs the remains of animals suggest a tribal totem; many ancient cultures recognised important power animals, an idea that has come down to us in much weaker form as the mascot that represents sports teams. In ancient times totem animals may have been invoked in spirit form to guide the tribal Elders when making decisions on behalf of the community. This shamanic technique is used by animist religions the world over, and animal guides are still invoked by Witches in trance states to take the seeker to a place where information is imparted.

# The Tomb of the Eagles, South Ronaldsay, Orkney

The Tomb of the Eagles is a ritual site at the heart of the Selkie shores. Dating from about 2400 BC, it stands close to the jagged edge of the cliffs overlooking the Atlantic and appears as a grassy mound with a narrow entranceway. From the cliffs, seals lounge and play and have been the

guardians of the tomb since it was built. The chambered tomb, also known as the Ibister Chambered Tomb, is home to many dead whose skulls were placed around the edge of the interior chambers. From the entrance, the sea stretches far out to the horizon.

The tomb is so-called because the remains of white-tailed eagles have been discovered inside it. Some of the birds' feet are placed alongside necklaces and bones within the tomb in a way that suggests the tribal totem. In two other tombs on the islands there are dog remains; at another there are songbirds and at still another, cormorants. The white eagles, once common to the area, were hunted to near extinction and have since had to be reintroduced to Britain.

South Ronaldsay is an enchanting place and all the more so because of the warm reception we received from John Hedges, who excavated the tomb, and his family. They encourage visitors to handle the ritual tools and the skulls that have been recovered from this site. One of John's daughters spoke to us in the soft lilt of the islands, amazement shining on her face as she told us tales of the ancestors' achievements. We held the skulls and she showed me the hereditary brain anomalies on a female skull, which accounted for a misshapen head.

Holding ancient ritual objects is like touching the rituals themselves. The small ritual axe head was so finely polished that it had the appearance of a piece of modern manufacturing. A jet bead looked like shiny plastic, yet when I touched it, its age and handmade feel were unmistakable.

Details of this small Neolithic community are easy to establish because so many skeletal remains have been found in the tomb. The men averaged 5 feet 7 inches in height and the women 5 feet 4 inches. Their average age was between 15 and 30, although some lived to be 50. The size of their skulls suggests they would have been as intelligent as we are, and they ate a healthy diet of seaweed, fish, eggs, lamb, beef and, of course, no sugar. Some of the skeletons showed signs of degenerative diseases such as osteoarthritis, but what interested me more than anything else was that none of the people here seemed to have died from any act of physical violence.

There is a Bronze Age house nearby which gives clues as to how meat was cooked. A stream flowed into the house into a large stone basin in the centre of the small flagged stone floor. Here, the water was contained and boiled by heating stones. John Hedges has recreated this cooking technique in the house: it took three hours to heat the fire to fuel the stones; three hours to boil the water in the basin; and a further three hours to cook a sheep. The heated flooring and the little cupboards are charming. It is the kind of experience that makes it possible to imagine what it would be like to have strange visitors from another time suddenly dropped into your own living room. The details of ordinary life are often more fascinating to me than the history of kings and queens. Here it is possible to sit with the ancestors outside of ritual space.

**THE PASSAGE TOMBS AT KNOWTH**

*(opposite) and Dowth in County Meath, Eire, are spectacular both in size and in the magnificence of the megalithic art displayed inside the chambers and on the outside of the monuments.*

To enter the tomb itself, one has to lie flat on a large skateboard. By lying on your back and holding the rungs above, you are able to wheel yourself into the tomb. It seems strange to enter a place like this on wheels. Once inside, the tomb is a quiet place; the silence of the ages descends. There are many chambers, all once filled with skulls, bones and beads. The skulls were placed around the interior walls of the chambers, staring out towards anyone entering the tomb. It is empty now except for a few skulls in one chamber to help the visitor imagine how it once looked.

# Newgrange, Boyne Valley, Eire

In this valley is the largest concentration of megalithic art in the British Isles. Newgrange and the nearby sites of Knowth and Dowth are the most sophisticated examples of chambered passage tombs in Europe. At Newgrange the Earth Goddess is represented by a large mound into which a shaft of Midwinter sunlight penetrates the heart of the womb. The Midwinter solar alignment means that for five days around the time of the Solstice the sun casts light through a small roof-box that illuminates the interior passage and then creeps along until it reaches the central chamber inside the tomb. The three interior chambers are placed much in the way a Neolithic Goddess would have lain in wait for her lover. The chamber at the centre is the heart of the belly. The furthest chamber forms the head into which the light is received. To

each side of the central chamber are two adjacent chambers, both of which housed the cremated remains of dead ancestors. The ancient rites enacted in the womb-tomb were observed as the golden ray of light shone gradually through the birth passage to illuminate the mystery at the centre. Inside each chamber is a basin stone smoothed by time and by the rolling of stones on its surface. Quartz has been found resting in the basin stone, where it may have been placed waiting for the sun to illuminate it.

Newgrange's exterior walls are also covered in quartz. The walls have been reconstructed according to the patterns in which the fallen stones lay and archaeologists have placed them the way they may have been 5000 years ago. The fact that this temple would have shone startling white against the surrounding landscape, and that quartz is much used in crystal magic, means that many ideas have been put forward as to the significance of the stone. When quartz is struck it gives off sparks, and since the method of 'napping' flint was widely used to create tools in Neolithic times, it would have been common knowledge that the striking of two quartz stones created a wondrous sight of sparks dancing in the air and must have been held in some esteem.

Of course this is only conjecture. We cannot say more than that quartz has been held as a sacred stone in many cultures and in oral traditions is known for its healing properties. But it is surely not insignificant that the entire mound of Newgrange is covered in quartz and that quartz was also found inside the tombs in what seems to be ritual space.

The recurring motifs of megalithic European art found at Newgrange are some of the most fantastic

**THE MAJESTY OF NEWGRANGE** *as Earth Mother. The light shines in through the roof box above the door. Insets, spiral art representing the cycles of the Goddess, our movement through life in all its phases to death and the afterlife.*

in Europe. There are spirals, triangular and lozenge-shaped diamonds, cups and rings, and circles with radial lines emanating from the circles. They are empty signs that resonate with our spiritual unconsciousness. We look at them with eyes shaped by our history and context. There are no absolutes here.

There are many ideas as to who the artists were. Some believe a cult of artists existed within the community. Some archaeologists maintain that the spirals were created during rituals, with the experiences of a drug-induced trance to aid the artist's vision; the artists may have been the elders who enacted ceremonies to mark the yearly cycle. There are those who believe the diagrams are crude illustrations of the tomb and its surrounding landscape and the wavy diamonds are the waters of the River Boyne. What appear to be sun and flower motifs could just as easily be stars. There are also many reasons to believe that these are images of the Great Earth Goddess, the triangles being representations of the triangle found between women's legs. Marija Gimbrutas suggests that the diamond shape is a reference to the Great Goddess and the spiral to the Eye Goddess originally found in Mesopotamia. The spirals and radiating eye shapes were later made into intricate patterns with the influx of the Celtic culture. They resonate with the image of the maze, which is found all the way from Crete to Cornwall. These early carvings may be perceived as gateways to the lives of the ancestors. Sacred art such as this, which strives to communicate the experience of ritual, leaves us with the conviction that something very moving and profound must have gone on to produce such lucid imaginings on rock.

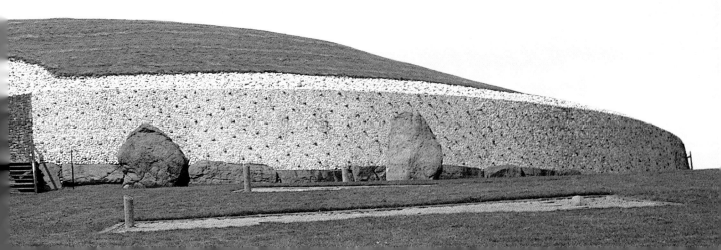

## Loughcrew Passage Tomb, Westmeath, Eire

My first encounter with the Loughcrew passage tomb was in the spring of 1999. I had heard about the extraordinary Neolithic art inside it, and also the fact that you were able to gain access by asking a nearby farmer for the key. The tomb is part of a sacred landscape dated to approximately 3500 BC. In 1828 the antiquarian Louisa Beaufort wrote: *In the county of Westmeath, in one of the Hills of Loughcrew, which are called by the peasants the Witches Hops, is an extensive excavation, consisting of three large chambers with a narrow passage leading to them. In one of these rooms is a flat altar-stone of considerable size; near to this artificial cave stand two lofty pillar stones known among the people by the names of 'the Speaking Stones' and 'the Whisperers'. Names evidently traditional of there having been oracles or divinations given from these 'dark places of the earth'.*[1]

**LOUGHCREW** – *the view from the chambered tomb (below). Opposite, the mound of Loughcrew in the evening light. The tombs are said to be 3500 years old, and the individual chambers spoke very profoundly to Jan and me.*

On a warm spring evening I walked up the hill with my friend Jan. We came equipped with incense, candles and Tibetan bowls to accompany the spiritual work we intended to do. We entered the tomb with a sense of expectancy, but once we were in the passage and lit a candle, we were astonished to see the walls of the inner chambers come alive with some of the most beautiful Neolithic art I have ever seen. Images of flowers, geometric designs and circles with lines radiating outwards created an inner sanctum of great beauty. The central passage into the rounded mound opened up into three antechambers. I was struck by the middle chamber, which was adorned with images of flowers and spirals. There was a kind of warmth here that spoke to me and invited me in. Jan, on the other hand, felt compelled to enter the chamber on the right, a dark place that seemed unwelcomingly still and heavy to me.

Often I am struck by how a place can speak to one person and not another. In this case we both had an immediate reaction to the difference in the three chambers. In the central chamber we lit a candle and incense. Then, wanting to go into an antechamber to do some chanting and meditative work, we both spontaneously entered the one that had chosen us.

Settling down into the dark space, I noticed that the chamber on the left – which had appealed to neither of us – seemed to be the most active of the three. It was here that the light from the passage shone in and it was also the chamber into which the incense blew. The central chamber, where I sat, was warm and comforting. Even the stone felt warm and the beautiful images of flowers were joyous. Jan's chamber was colder, still and dark.

As we chanted and spent time playing the bowls, the eerie sounds circulated around us and we both experienced very profound feelings. Jan had recently been suffering the pains associated with a difficult and early menopause. She had tried Hormone Replacement Therapy and various alternative therapies, but nothing seemed to work. Yet in this dark place she emerged content with the issues raised by the menopause. This is often a difficult time for women, as they undergo a devaluing of their power in patriarchal representation, but it is also the time of the Crone and the imparting of that wisdom that comes with age.

**AN OGHAM STONE** *grave marker on the Dingle Peninsula (top). Above, cup and ring spiral art, Dingle. Right, cairn at Maeve's Seat, County Sligo. Maeve was a powerful and sexually voracious queen in Irish legend, also commemorated at the Hag's Chair. Opposite, Tom Lux, a Californian Pagan, wishing at the Hag's Chair.*

I, on the other hand, experienced a strong desire to be a mother. I had always thought that having children was something I would do when the time was right, but as the years went by for one reason or another the time never was right. Yet here in this sacred womb tomb I realised how important it was for me.

I looked at the structure of the passage and understood the analogy that this was the birth canal, lit up once a year with the carefully calculated movement of the sun. This chamber is aligned to the Autumn Equinox, on which date it is penetrated with a shaft of light. But it also occurred to me that here we had a representation of the Triple Form Goddess with her three stages of womanhood, all equally important and valued. I had chosen the Mother chamber and Jan the Crone chamber; the light, active Maid chamber had not been relevant to either of us at the time. The knowledge imparted to us during our meditation was directly concerned with Feminine imagery.

IN HER EXCELLENT DESCRIPTION *of the cairns and the mythology surrounding Loughcrew, Jean McMann mentions several other female legends. Clearly this is a site of traditional Goddess Worship. The Caillaigh was known to* **harvest a field faster than any man, [was] a crone lamenting her youth, a banshee announcing a death, or a Christian nun.** [3]

*The regular occurrence of female characters in these stories suggests that this was a potent Goddess site. The amazing experiences I have had there, all of which spoke of woman-centred spirituality, make me consider this a power spot for women and for men seeking the Goddess.*

# The Hag's Chair

Loughcrew also houses the Hag's Chair, a single stone that forms the base of the tomb and is recognisable as a massive seat that faces north. It is also known as the Seat of Visions, for it is said that if you sit on it and gaze out over the landscape you will see visions. The Oldcastle postman, Joseph Carroll, told this story in 1948.

*The 'Hag's Chair' is up at the middle hill – Slieve na Caillaigh. The old tradition is that Queen Tailte or Queen Maeve sat in it and proclaimed the laws that were to be observed by the people. There is another old tradition that an old hag called the 'Caillaigh Waura' used to sit in it, and it was from that they called it 'The Hag's Chair'.*

*Another legend about her is that if she'd carry her apron filled with stones and jumped from each of the three hills to the other, she'd be mistress over the whole of Ireland.*

*She started from the Cairne Bawn, and when she was jumping she dropped a handful of stones that accumulated to hundreds of tons. She hopped to the next hill, Slieve na Caillaigh, and she dropped another handful of stones there, and it accumulated into a great cairne. She jumped to the third hill – I forget now what they call it – and threw the remainder of the apron of stones there. And they turned into another great pile, and the three piles of stones are there to the present day. She was going to jump from the last to a hill at Patrickstown, and fell and broke her neck.* [2]

Traditionally one can make a wish at the Hag's Chair and it will be granted. So I was not entirely surprised that on one of my later visits to Loughcrew I experienced first-hand some Pagans doing just that. The day before I was due to photograph the chair I had been at Newgrange, and as I came out of the site at sunset, I saw a woman hitchhiking on the road out of the Bru Na Boyne visitor centre. She struck me as a rather hilarious sight, pulling a huge suitcase on wheels and wearing a wide-brimmed hat that made her look like a cowgirl. I offered her a lift, took her part of her journey and wished her well.

The next day at Loughcrew there was the same woman with the hat, though this time without the suitcase. We recognised each other and got to talking; she asked me if I was a Pagan. I said that I was, and we talked as we went up the hill. She introduced me to the other travellers with her. One of them was Gavin Bone, a Witch and writer on Pagan lore of whom I had heard through my connections with the London Pagan scene. I was amazed when Gavin told me that he lived with Janet and Stewart Farrar, the two most famous Witches in Ireland, not far from Loughcrew. I had been to Ireland many times and each time I went, Doreen Valiente asked me if I would visit the Farrars. Each time I said I didn't know them and each time Doreen said it didn't matter; that as a friend of hers I would be welcome. Time and again at sacred sites I have experienced a weird synchronicity in which I meet people who are to be important connections. Doreen had told me that the Farrars and Gavin were people I would connect with in the spiritual work I am doing, but I had no idea that we would get along so well. Many a time I have met so-called 'important' Witches and not had a resonance with them. This time there was immediate recognition.

*AT THE HAG'S CHAIR* *the wishing is done by placing your hands in two indentations on each side of the chair and making your wish silently. The chair is also called the Seat of Visions, as it is said that if you look out over the landscape you will see the future.*

What was so incredible about this meeting was that I had not intended to be at Loughcrew at that hour – it was only the heat of the day that caused me to delay my visit and to arrive at precisely the same time as Gavin.

The other members of the party were two practising Pagans from America, Tom and Val, and Diane, the editor of the Pagan magazine *Pangaia*. Tom and Val had brought an instant altar that Tom placed on the Hag's Chair. It consisted of a

transparent tube that disconnected into four compartments, each of a different colour and with something to represent one of the elements. In the east Tom placed a little box with a feather in it and it was yellow; in the south, cayenne pepper and it was red; in the west a shell and it was blue; and in the north some earth and it was green. These represented Air, Fire, Water and Earth respectively. In the centre Diane placed her Celtic cross. We ate and drank at the base of the chair. Gavin made a libation to the Goddess by pouring a small bottle of red wine on to the stone and we left a piece of bread on the chair itself. These acts of devotion are ancient rites to acknowledge the sacred space with an offering.

After we had spent this time outside, we entered the tomb and Diane mentioned that the right-hand chamber, the Crone chamber, had a simple but decipherable image of tea-making on the walls: a leaf drawn over the image of a container and then the container as a full vessel. It seems only fitting that someone all those thousands of years ago should have made tea for the Crone.

Once we emerged from the tomb, I felt the desire to sit on the chair itself. I projected my energy across the land and felt as if I was seeing its ley lines, as if for an instant the veil lifted and I could see the magic of the land. I had been in Ireland for two weeks and this was my last evening. In the course of my journey I had had word that Doreen Valiente was not well. With these fellow healers, I took the opportunity to project healing energy to her from the Hag's Chair, standing with hands held around the seat.

It was Lammas the next night and I returned to England for a rite with my coven. As I celebrated the ritual of the land and the cutting of the harvest I asked my group to send healing energy to Doreen. When I spoke to her the next day she told me that she had inoperable cancer and that this would be the last Lammas she celebrated. She laughed at the timing and said, 'Happy Lammas' with a wry giggle. It somehow seemed right that, after she had asked me so often to visit Janet and Stewart Farrar, I should finally meet them now.

*SPIRAL CARVINGS (left and far left) which evoke memories of the ancestors who produced them. Above, diamond shapes are another traditional form of engraving – the sexuality of the Goddess is in the triangle.*

Doreen spoke of the magic of Lammas, how it is important to know when to cut the harvest and to reap the best of the crop. The symbolism of this seasonal festival is all about the land which is ripe for harvest. The Lammas King traditionally gives his life willingly at the height of his potency rather than maintain his leadership after his strength has waned. Doreen understood that she was leaving this world, not by going into a decline or wane, but at the height of her potency. She died a month later, on 1 September 1999.

The journey I took to many of the sacred sites described in this book was with a heart made heavy by the loss of a great friend, but the magic of the timing was extraordinary. Doreen was a great Elder who taught me many things on my inner journey. In the course of writing about sacred sites I had her as one of my inspirations and it seemed fitting that my journey paralleled her last journey. As she was leaving she somehow sent me to others who would become friends and this I consider to be no accident. My visit to the Hag's Chair enabled me to have a psychic connection to my Crone teacher and she left me with the gift of new Witch friends whom she also loved.

# AFTERWORD: RETURN TO AVEBURY

Avebury is a village inside a sacred circle and as such is always open to visitors. On the day Doreen Valiente died, I visited West Kennet long barrow, a place full of the imagery of the Crone. In the central chamber I lit a candle and some incense and placed a crystal she had given me in a nook in the wall. As the chamber filled with incense and the glow of the candle, I said my farewells to my friend. Some visitors passing through commented more on the beauty of the candle flickering in the heart of the tomb and how its light evoked ritual, than on the fact that I was obviously using the place for meditation. I was able to continue with my private ceremony and, because there was enough room, it did not stop them from seeing the interior.

In Avebury, there are many instances of people taking ritual to the stones. I have witnessed stone-hugging sessions and groups of dowsers wandering around with copper rods. On the site of the old maypole post-hole, which is at the heart of the fertility imagery, flowers are often left as an offering. Petals are strewn on stones and wine poured as a libation. It is also common to see visitors sitting in the Devil's Chair. Throughout the year, whenever there are seasonal festivities, there are people doing strange and wonderful things. The excellent Stones Restaurant and the Henge Bookshop make

Avebury a place to go no matter what the weather. Here there are sources of information on many Neolithic sites, folklore and Paganism. The reading material is so comprehensive that even if the visitor is a newcomer to Pagan imagery it is easy to access the information and wander the sites in an informed way.

Pagans are abundant in Avebury, often made obvious by their colourful attire or the staffs carried by wanderers. Once I met a man in robes who had walked barefoot from Stonehenge on the Summer Solstice. He was not the only one.

Despite the coaches full of tourist couples in matching outfits, nothing can spoil the beauty of the circle or the awesome sight of a henge dug by hand. It is truly monumental in scale. The continual recurrence of crop circles in the adjoining fields is often seen as a sign of the magical qualities of the stones. This is taken as a matter of course. Nobody bothers much about the debate over whether these are hoaxes or alien signatures – least of all the farmers, who cut the corn no matter whether they are destroying proof that there is life in other galaxies.

Knowing that the place is so magical and has been considered so for centuries, when I go there I use the old Celtic method of divination. This is simply to ask with the heart for a sign of what I need to know. As I wander among the stones I look for visible clues to my state of mind. On the day Doreen died, I was due to go to Avebury anyway. As I went I watched for signs that would augur her passing. After meditating at West Kennet I went to the central circle. As always there were some visitors, but as I walked to a more private area I noticed two crows sitting on the stone I was heading towards. Knowing that crows are the harbingers of death in Celtic mythology I watched these two dance on the top of the stone. Then one flew to the next stone, to be followed by the other. As I walked around the circle the birds flew with me, always one stone ahead. I had never seen birds dancing like this on the stones and immediately thought that this was one of the signs I was looking for. With one bird always a little ahead of the other, then briefly catching up, I allowed my thoughts to wander and connected to Doreen's meeting with her beloved

**AN UNUSUAL VIEW OF SILBURY HILL**
*(previous page). Opposite, West Kennet long barrow, a mysterious ritual site where I said goodbye to Doreen Valiente. Below, the crows, a Celtic symbol of death and afterlife, sit on a stone at Avebury: Ron and Doreen reunited.*

partner, Ron Cooke, who had passed over two years earlier. I held the thought that this was a symbol of reunion with the beloved and felt that she was happy.

At the end of the day, as I left the circle, I saw the largest migration of birds that I have ever seen. They came from one horizon and flew all the way to the other. It was a spectacular sight and I found myself thinking that Doreen had chosen to migrate with the birds. It was a beautiful image and a way to access those power animals that come to us when something cannot be spoken, just felt. Doreen had flown with the birds.

This is one way to use the ancient sites for important personal thresholds. The last time I went to Avebury it was my birthday. I was wanting nothing more than a day out from London, and it was made that much more special by going to a sacred site. I sat in the Devil's Chair, walked around the henge and then had a birthday drink in the pub by the side of the old well. The well is a feature of the pub, for it has been made into a table by covering the hole with thick glass. I looked down the well and, in the ways of old, made a wish.

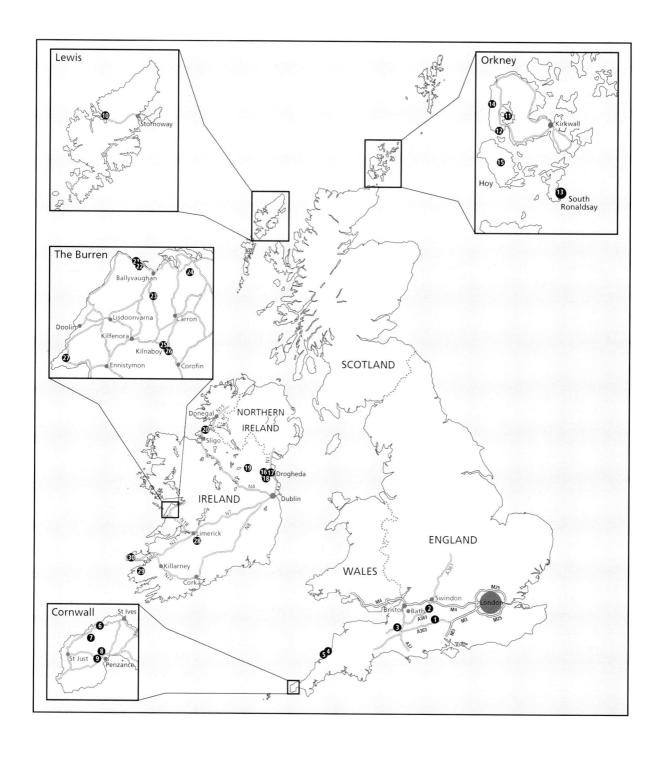

# How to get there

The sites in this book are treasures and in order to preserve the ever-declining freedom to wander in sacred sites it is important that Pagan visitors honour  the  sites and the land. Please do not leave litter and, as an extra gesture to the Earth Goddess, pick up any litter you find and take it to a rubbish bin. Although the burning of incense and candles makes for a great rite, please be sure that candles do not burn  unattended.  Be nice to local farmers and they will continue to allow us access to their land. Do not disturb birds' nests or wild animal habitats. As someone wise once said, 'She who clears up the garbage from her path will walk without obstacles in her way.'

## SOUTHERN ENGLAND

1 **Stonehenge**  is visible from the A303, about 15 miles west of Andover and 50 miles south-west of London.

2 **Avebury** is on the A361, 12 miles south of Swindon and 50 miles west of London

3 **Glastonbury** is on the A361, about 25 miles south of Bristol and 100 miles west of London.

In Glastonbury, the Goddess and Green Man Bookshop is not to be missed: in addition to covering Goddess and Pagan literature, it also stocks statuettes of the Goddess, jewellery and Pagan artefacts:
2–4 High Street
Glastonbury
Somerset
BA6 9DU
Tel: 01458 834697
email: goddess.shop@uk.com
www.goddessandgreenman.co.uk

The Annual Goddess Conference is organised by Kathy Jones and Tyna Redpath. For information and prices, contact:
The Avalon Foundation
2–4 High Street
Glastonbury
Somerset
BA6 8DU
Tel: 01458 833 933 or 831 518
Fax: 01458 831 324
email: Kathy.Jones@ukonline.co.uk

## CORNWALL

Useful information is available from:
Cornish Tourism
01872 322 900

4 **Boscastle**, home of the Witchcraft Museum, is on the Tintagel road, a right turn off the A39 between Bude and Wadebridge.

The Museum of Witchcraft
Graham King
Boscastle
Cornwall
PL35 0AE

email: museumofwitchcraft
@compuserve.com
www.museumofwitchcraft.com

5 **St Nectan's Kieve & Tintagel Castle** (legendary birthplace of King Arthur and home to Merlin
Tintagel
Cornwall PL34 0HE
Tel: 01840 770328

Opening times: April–Oct 10 a.m.– 6 p.m.; Nov–March 10 a.m – 4 p.m. Closed 24 – 26 Dec.

6 **Zennor** is on the B3306, about 10 miles south west of St Ives.

7 **Men-an-Tol Holed Stone**: south-west of Madron (reference 426 349 on the Ordnance Survey map). Take the A30 from Penzance towards Lands End and turn right at the B3312. Men-an-Tol is marked off to the right – follow a clear and easy track for 20 minutes, then turn off to the right to the site. Continuing along the track will bring you to the Nine Maidens Stone Circle.

8  **Madron's Well**: *take the A30 from Penzance, then turn right on the B3312 to Madron. The well is marked on the road – a 10-minute walk along an easy path.*

**Nine Maidens Stone Circle (Boskednan):** *follow the directions for Men-an-Tol above, then continue along the track for a further 5-10 minutes. The stone circle is clearly marked and is also on OS maps. It is easy walking, but be prepared for a round trip of at least 40 minutes.*

9  **Boscawen-noon Stone Circle:** *take the A30 from Penzance toward Land's End. Go past Drift, then past the turn-off to St Buryan. 1 mile beyond the turn-off is a small layby with a kissing gate – if you get to Crows an Wra (Witches Cross) you have gone too far. Go through the gate and follow the overgrown path to the circle. Consult a local OS map for directions as the circle is not far but often hidden by gorse. It is, however, always possible to get into it. This walk is not recommended for disabled visitors or those with problems walking on rough terrain.*

SCOTLAND

10  **Callanish**  *The Isle of Lewis is reached by a ferry from Ullapool to Stornoway. From Stornoway, take the road towards Harris and turn right towards Achmore. The road goes straight to Callanish, which is very well signposted. The Visitor Centre has exhibits, shop, toilets and a cafe, as well as information about other sites in the area. Margaret and Ron Curtis's number is available at the centre, for those who want a guided tour.*

*Callanish Visitor Centre is open all year round – check for times.*

*Tel: 01851 621422*
*Fax: 01851 621446*

*Other useful information is available from:*

*Lewis Tourist Office*
*26 Cromwell St*
*Stornoway*
*Isle of Lewis HS1 2DD*
*Tel: 01851 703088*
*Fax: 01851 705244*

*Scottish Tourist Board*
*Tel: 08705 511511*
*www.holiday.scotland.net*

*British Airways (the main airline to the Western Isles and Orkney)*
*Tel: 08457 222111*

*Scotrail*
*Tel: 08457 550033*

*National Rail Enquiries*
*08457 484950*

11-15  **Orkney**
11  **Ring of Broadger and Stones of Stenness**
12  **Maes Howe**
13  **Tomb of the Eagles**
14  **Skara Brae**
15  **Dwarfie Stane**
*All these sites are on the Orkney Mainland, except the Dwarfie Stane, which is on Hoy.*

*Information is available on http://www.orkney.com (emails to Tourism Orkney: info@otb.ossian.net)*

*Orkney Tourist Board (and the Orkney Archaeological Trust)*
*6 Broad Street*
*Kirkwall*
*Orkney KW15 1NX*

*Tel: 01856 872856*
*Fax: 01856 875056*

*Stromness Tourist Office*
*The Ferry Terminal Building*
*Pier Head*
*Stromness*
*Tel: 01856  850716*

*Scara Brae Visitor Centre*
*Skaill*
*Sandwick*
*Tel: 01856 841815*

*Maes Howe*
*Torminston Mill*
*Stenness*
*Tel: 01856 761606*

*For access to the Tomb of the Eagles, contact:*
*Mr & Mrs Simison Liddle*

South Parish
South Ronaldsay
Tel: 01856 831339

Kirkwall Airport: British Airways
reservations tel: 0345 222111

Flights from: Aberdeen, Glasgow,
Edinburgh, Inverness

Connections from: London
Heathrow, Birmingham and
Manchester

Rail Enquiries: from overseas:
44 141 335 4263

IRELAND

16  **Newgrange** – take the N2 out of
Dublin north towards Slane.
Signposts for the centre are clear,
to the right after about 40
minutes.

17  **Dowth** is about a mile east of
Newgrange on the same road.

18  **Knowth** is 3 miles south-west,
on the road between Drogheda
and Navan.

For more information contact:
Bru-na-Boyne Visitors' Centre
Donore
Co Meath
Tel: 353 41 988 0300 or 982 4488
Fax: 353 41 982 4798
Opening times: Oct-Feb 9.30
a.m.–5.30 p.m.; March-April 9.30
a.m.–5 p.m.; May 9 a.m.–6.30 p.m.

19  **The Hag's Chair and
Loughcrew** are on the R195 to the
north-west of Kells (Ceanannus
Mór). From Drogheda, follow the
N51 to Kells,then bear right on the
N3 towards Cavan (An Cobham).
Loughcrew Megalithic Cemetery is
signposted off to the left after
about 12 miles.

20  **Crevykeil** is to the right off the
N15, about 10 miles north of Sligo.

21-27  **The Burren**
  21 **Well of Holy Cross**
  22 **Best tea on the Burren**
  23 **Toothache wells**
  24 **St Colman's Well**
  25 **Innerwee's Backache Chair**
  26 **Kilnaboy**
  27 **St Bridgit's Well**

Maps and information about the
many wells on the Burren are
available from:

The Burren Visitor Centre
Burren Exposure, Whitethorm
Ballyvaughan
Co. Clare
Tel: 353 65 77277
Fax: 353 65 77278

28  **Lough Gur** is on the R514 about
12 miles south of Limerick.

29  **The Ring of Kerry** is on the N70,
about 30 miles west of Killarney.

30-31  **Dingle:** Information on the
prayer chambers of the Dingle

Peninsula may be obtained from:

www.dingle-peninsula.ie.
Celtic & Prehistoric Museum
Dingle Peninsula
Tel: 353 66 915-9941
Celtmuseum@hotmail.com

Archaeological tours are available
through:

Sciuird Archaeological Adventures
Holyground
Dingle
Co Kerry
Tel/fax: 353 66 915 1937
Email: arch@iol.ie

In the USA, for information on
archaeology in Dingle:

Tel 510 472 6325 and ask for Laurie
McNeill.

Other useful information is
available from:

Irish Tourist Board
150 New Bond St
London W1Y 0AQ
Tel: 020 7493 3201
Fax: 020 7493 9065

Dublin Tourism Centre
Tel: 353-1-605-7777
Fax: 353-1-605-7787
www.visit.ie/dublin
www.visitdublin.com

reservations@dublintourism.ie

Aer Lingus
From Britain 0645 737747

# Glossary of terms

**Casting a circle**   The act of tracing a circle to designate an area of sacred space in which a ritual will take place. The circle is traced by sending energy through the athame {Witch's blade), wand or directly through the fingers. Energy generated in the solar plexus is sent out through the fingers and forms a protection and a barrier for the Witch to work within. For indoor working it is not possible to engrave the circle on the floor, so the circle is figurative, not literal. Any ritual work within the circle is considered to be 'between the worlds', a place where transformation takes place.

**Centring**   or grounding is achieved by various methods of relaxation, meditation or using breath work to calm the mind as preparation for a ritual. It is done before a rite, but is also used to calm or balance the mind so that a person may meditate more effectively.

**Chakras**   The seven key energy centres of the body according to Hindu Yogi sources. They are imagined as wheels that spin and are connected to the function of certain organs, as well as to states of mind. Though there are seven key chakras, there are many 'lesser' energy centres. Each corresponds to a different colour and can be focused on in meditation by visualising that colour at the correct position for the particular chakra. Chakra cleansing is part of meditation that uses breath to connect to that chakra, imagines a colour and clears the chakra of 'stuck' energy.

**Charging**   A traditional method used by Witches to energise an object. It involves imagining energy as a colour, essence or electrical pulse and sending it into an object that may be used for healing.

**Deosil**   In a sunwise or clockwise direction. In Witches' rites the use of direction is understood to be either a gathering or a dissipation of energy. Deosil is the direction of gathering, generating, increasing and positive accumulation. Circles are cast in a deosil direction to gather the energy of the place and people for a rite. To end a rite, certain kinds of working involve the banishment of the circle widdershins or moonwise (anti-clockwise), to expel the energy of the rite out into the universe and close the circle so that there are no traces left of what has taken place. In modern parlance, it is used to clean up the energies of a rite.

**Druidry**   One of the oldest Pagan religions in Britain. Mentioned in the chronicles of Julius Caesar, the Druids were the old priests of the Britons. Modern Druids are famous for their worship at stone circles and in particular at Stonehenge, where the Order of Bards and Ovates conducts rites at the Summer and Winter Solstices. They are particularly connected to sun worship and solar events.

**Elements**   In Wicca, the name given to Air, Fire, Water, Earth and Spirit or Aether. The four key elements are used as a means of balancing in meditation or in a ritual where the fifth, Aether, is then invoked.
Each element is connected to a direction. East is imagined as the Element of Air and is concerned with issues of the mind, ideas, the Sky Father, the word, communication, writing, laughter, music and breathing. The element of Air is invoked by facing the east and imagining a symbol of Air, which for many is a bird – particularly an eagle – or the sky on a clear day. It is the element represented by incense or the athame.
Fire is the element of the south and rules all issues of passion, lust, the phallus, issues of the Masculine, inspiration, healing through warmth or purging. It is imagined as a hot desert, a lion, flames or a volcano with fiery explosions and is represented by a candle or a wooden wand.
Water is the direction of the west and governs issues of love, flowing communication, emotion, the depths of the personality, the journey of the soul, the womb, birth, the Feminine and voyages. When invoked as part of the directions it is imagined as a great body of water, a dolphin, whale, fish or mermaid. It is represented by a cauldron, cup or chalice filled with water.
Earth is the direction of the north and refers to the material realm. It governs issues of the home, family, money, work, endurance, steadfastness, commitment, silence, inner power, the Earth, environment, crystals, stones and the manifestation of ideas. It is represented by the pentacle, a disc of copper or wood that has the

symbols of Witchcraft on it. It is also imagined as a crystal, a stone or some earth.

Aether is more than the sum of the four and is governed by the idea of 'to go'. It sits at the centre of the circle where the magic occurs. Magic is always an issue of transformation and therefore falls within the domain of Spirit.

**Female mystery**  An attempt to describe a rite that refers to the female experience. The idea of a female mystery is in part a reference to Goddess worship. Within Wicca there is a reverence for the Goddess and for the idea of the Feminine. Goddess mysteries have in them direct references to birth, menstruation, menopause and the image of a deity as female. They are not, however, wholly dependant on women participating at a female ritual. If a rite has within it a birth reference it can be said to be a female mystery, but this does not mean that it excludes men, or women who have not given birth

**Grounding – see Centring**

**Healing**  There are many ways to perform a healing, but the term refers to anything that has the intent of healing. In some instances a Witch may decide to send some distant energy to a person who is not well or to hold the hands over a wound and send energy into the injury to heal it. Other methods include 'hands on'; sending healing indirectly via a crystal; or actively using the techniques of visualisation. The most important aspect of a healing is the intent.

**Ley Lines**  Energy lines that run through and across the land as the meridians in acupuncture run through the body. They are not normally perceivable by sight (though some people say they can see them), but are understood as the linking of sacred and spiritual sites to natural power centres on the Earth's surface. The ancient temples of stone, cairns and holy wells are believed to be linked via ley lines. After the introduction of the Christian faith into Britain, the Church built upon these ancient energy centres. The energy, however, does not change and the Pagan sites are pointers to the special places that held spiritual importance to the ancestors.

**Mystery Religion**  One of many religions, including Wicca (see below), in which the teaching involves a hidden element or initiation that is not made public. Mystery religions keep certain teachings for certain rites, times and places. The mystery is not discussed, but relates to the process of transformation that takes place within the consciousness of the initiate, who cannot then put that experience into language. It may also refer specifically to the oaths that bind the participant from telling those uninitiated to the mystery what is to occur. In this case it serves as a hidden teaching and also brings the element of surprise to a ritual.

**Pagan**  is a term used by Christians to describe those not of the Christian faith. It originally meant rustic or country dweller and referred to the indigenous people of Britain who

were considered to have a spiritual path in 'nature' religions.

**Pathworking**  A method of entering into a dream state or trance in order to follow a guided visualisation. This may be done in several ways, but is often spoken by a priestess or Witch to take others on a journey of inner discovery. The term can also be interpreted literally, using a walk to draw on the symbolism presented to the person who is journeying along a path. Whether the path is visualised or actually walked, the images, plants and animals encountered along it are guides to finding some inner knowledge. Two examples of pathworkings are given on pages 44 and 106.

**Shamanism**  The primal ground from which all spiritual traditions have emerged. It is the ancient religion of our ancestors, who took nature as their spiritual teacher. The shaman is the priest or priestess who conducts rites of passage for the community.

**Scrying**  A particular form of divination – that of looking into a mirror or crystal ball in order to gain knowledge or wisdom of events, people and places. The Witch's black mirror is used to see not outward but inner appearances, to gain insight into a situation by clearing the mind of any interference and allowing the gaze to be hazy. After some time images may appear on the inner screen of the mind or behind the third eye, located between the eyes on the forehead.

The word is also used to describe a method of allowing the mind to gaze without focus while 'intending' to seek an answer. Thus, for example, it is possible to scry for a book when asking the question, 'Which book do I need to read now?' Information that comes into consciousness after allowing the mind to relax is perceived to be that which you need to know now.

**Tantra**  The ancient spiritual paths that incorporated polar sexuality into their teachings. In Hinduism, tantric yoga uses breath work and postures to reach a state of inner well-being while working with a partner of the opposite sex. It often includes sacred sex, which seeks not to achieve male orgasm but to keep the man aroused so that he attains greater states of consciousness. Women, on the other hand, are encouraged to have many orgasms, as this is considered part of the sacred arousal. Tantric teachings revere sexuality as sacred and divine, and men and women as sacred essences of the divine principle.

**Wicca – see also Witch**  The modern term for Witchcraft. It derives from the Anglo-Saxon word Wicce, meaning one who shapes or gives shape or is wise. The terms Witch and Wiccan are interchangeable, but because of the widespread tendency to deride Witches, many prefer to qualify the term or use Wiccan.

**Widdershins – see Deosil**

**Witch – see also Wicca**  One who follows the teachings of Witchcraft as a spiritual path. These words have been so denigrated over the centuries that it is now impossible to use them without qualification. To Wiccans, a Witch is a woman or man who worships the old religion, which includes a concept of the deity as both male and female as represented by the Horned God and the Goddess. In Witchcraft there is both a celebratory and a practical aspect to worship. Witches follow the moon cycles, the seasonal cycles and perform magic in relation to specific timings in nature. Thus they may worship the Goddess with a blessing ritual or work a healing for someone. Both aspects of the Craft are part of the work of a Witch.

# References

Publication details for the books quoted are given in the Further Reading section opposite.

**BRIEF INTRODUCTION TO PAGANISM**
*1* Doreen Valiente, revised many times over from early Pagan sources

**CHAPTER 1**
*1* Ian McNeil Cooke, Journey of the Stones
*2* Ibid
*3* Ibid
*4* Courtney, 1890, quoted ibid

**CHAPTER 2**
*1* Ron & Margaret Curtis, Callanish: Stones, Moon and Sacred Landscape
*2* Ibid
*3* Gerald & Margaret Ponting, New Light on the Stones of Callanish

**CHAPTER 4**
*1* Tom Muir, The Mermaid Bride and Other Orkney Folk Tales
*2* Ibid
*3* Ibid
*4* McNeil Cooke, op. cit.

**CHAPTER 5**
*1* Quoted in Paul Broadhurst, Secret Shrines
*2* Leila Doolan, The Book of the Burren

*3* John M Feehan, Secret Places of the Burren
*4* Paul Broadhurst, op. cit.

**CHAPTER 6**
*1* John Leland, quoted in Michael Dames, The Silbury Treasure

**CHAPTER 7**
*1* Geoffrey of Monmouth, Vitae merlini, ed. and trans. J. J. Parry (1925, University of Illinois Press)

**CHAPTER 8**
*1* Quoted in Jean McCann, Loughcrew: The Cairns
*2* Ibid
*3* Ibid

# Further reading

William Battersby, Knowth: 10 Ages (1999, Navan, Ireland)

Janet and Colin Bord, Earth Rites (1982, Granada)

Martin Brennan, The Stones of Time (1994, Inner Traditions International)

Paul Broadhurst, Secret Shrines (1988 self-published; 1991 Pendragon Press)

Aubrey Burl, Prehistoric Astronomy and Ritual (1983, Shire Publications, Buckinghamshire)

Anne Cameron, Daughters of Copper Woman (1981, The Press Gang Publishers, Canada; 1984, The Woman's Press, UK)

Mary Condren, The Serpent and the Goddess: Women, Religion and Power in Celtic Ireland (1989, Harper & Row, San Francisco)

Ian McNeil Cooke, Mermaid to Merrymaid: Journey to the Stones (1987, Men-an Tol Studio, Cornwall)

Vivianne Crowley, The Old Religion in the New Age (1989, Aquarian Press)

Ron& Margaret Curtis, Callanish: Stones, Moon and Sacred Landscape (1990, self-published, Callanish, Lewis)

Michael Dames, The Avebury Cycle (1977, Thames & Hudson)

Michael Dames, The Silbury Treasure: The Great Goddess Rediscovered (1976, Thames & Hudson)

Michael Dames, Mythic Ireland (1992, Thames & Hudson)

Walter Traill Dennison, Orkney Folklore

and Sea Legends (1995, Orkney Press, Kirkwall, Orkney)

Janet and Stewart Farrar, The Witches' Goddess: The Feminine Principle of Divinity (1987, Robert Hale)

John Feehan et al., The Book of the Burren (1991, Tír Eolas, Newtownlynch Kinvara, Co. Galway)

John M. Feehan, The Secret Places of the Burren (1987, Royal Carbery Books, Cork)

Miranda J. Green, Dictionary of Celtic Myth and Legend (1992, Thames & Hudson)

Ronald Hutton, Stations of the Sun (1996, Oxford University Press)

Evan John Jones and Doreen Valiente, Witchcraft: A Tradition Renewed (1990, Robert Hale)

Kathy Jones, The Goddess in Glastonbury (1996, Ariadne Publications, Glastonbury)

Eamonn P. Kelly, Sheela-na-Gigs: Origins and Functions (1996, The National Museum of Ireland, Dublin)

Jean McMann, Loughcrew: The Cairns (1993, After Hours Books, Oldcastle, Ireland)

Kenneth McNally, Standing Stones and Other Monuments of Early Ireland (Appletree Press, Belfast)

Caitlin and John Matthews, Ladies of the Lake (1992, Aquarian Press)

Brian Merriman, Cuirt an Mhean Oiche, The Midnight Court, translated by Patrick C. Power (original 1780; translation 1971 Mercier Press, Cork)

Tom Muir, The Mermaid Bride and other Orkney Folk Tales (1998, The Orcadian Ltd, Kirkwell Press, Orkney)

M. J. & C. O'Kelly, Lough Gur (1978, Claire O'Kelly, Ireland)

Sean P. O'Riordain, Antiquities of the Irish Countryside (1942, Cork University Press; 5th edition, 1995, Routledge)

Gerald & Margaret Ponting, New Light on the Stones of Callanish (1984, self published, Callanish, Lewis)

Starhawk, The Spiral Dance (10th anniversary edition 1989, Harper & Row, San Francisco)

Starhawk: Dreaming the Dark: Magic, Sex and Politics (1982, Beacon Press, Boston)

Doreen Valiente, An ABC of Witchcraft (1973, Robert Hale)

Doreen Valiente, Witchcraft for Tomorrow (1978, Robert Hale)

Doreen Valiente, The Rebirth of Witchcraft (1989, Robert Hale)

J. D. Wakefield, Legendary Landscapes: Secrets of Ancient Wiltshire Revealed (1999, Nod Press, Marlborough)

For information about Pagan activities in Britain and the magazine Pagan Dawn, contact:
The Pagan Federation
BM Box 7097
London
WC1N 3XX
www.paganfed.demon.co.uk
enquiries to
Secretary@paganfed.demon.co.uk

# Photographic acknowledgements

All photographs except those listed here were taken by the author.

page 11 (inset) Werner Forman Archive
page 16 (both) C. M. Dixon
page 17 Images Colour Library
page 38/9 Travelink/Barbara West
page 43 (top) © 2000 by WARA, Centro Camuno di Studi Preistorici, 25044 Capo di Ponte, Italy
page 81 (top) Fortean Picture Library/ Lars Thomas

page 85 Fortean Picture Library/ Dr Elmar R. Gruber
page 98 C. M. Dixon
page 127 © Crown copyright NMR
page 154/5 (main picture) Bord Fáilte – Irish Tourist Board
page 162 (main picture) Bord Fáilte – Irish Tourist Board
page 163 Bord Fáilte – Irish Tourist Board

# Index

Numbers in **bold** type indicate illustrations

animals, totem 150
apples 142
Arthur 137–9, 141–2
Avalon 137–42
Avebury **124**, **125**, **128**, **167**;
    landscape 55, 123–5, **128–9**;
    ley lines 132, 140; processional
    routes **131**, 131–3; rituals
    165–7; seasonal rites 124–5

Backache chair **110,** 110–13
Basques 130
Beltane 30, 57, 124, 125, 135
Berna Bridge 57
Bride 65–6, 99, 139; Bride doll **52**
Broadger, Ring of 79–80, 93
bulb ritual 72
Burren **102**, 102–5, 115, 116

Callanish: accounts of 55;
    avenues 131; circle shapes 53;
    Goddess figure 53–4; great
    circle **13**, **50**, **53**, **57**, **58**,
    58–62, **64,** 80; Maid, Mother
    and Crone **54**, 68–9, **71, 73,
    74**; moon-aligned 25, 54–5,
    58–9; rites 57; processional
    route **133**; smaller circles 63;
    stone shapes **67**, Weird Sisters
    **61**, White Cow Goddess 56
candles: blessing 37; Maid rite 73
Carnac 16, **38–39**
centring yourself 29
chakra system 59
Chalice Well **9**, **139**, 140, **140**,
    142, **142**, **143**, 143–4
Chanctonbury Ring 12, 75
clooties 116, **116**, 117
cord rite 35–7; for the Crone 71
corn dollies 52, **52**
Crone **144**; rites 70–1
cross-dressing 134
crystals 73–6; cleansing 77, **139**
cunt 130

Devil Stone **134**, 134–5

Devil's Chair 165, 167
Dingle Peninsula **14**, **19**
dowsing 11, 97, 132, 165
Dowth passage tomb126, 153
Druids 14, 30, 138
Dwarfie Stane **6**

Earth Goddess 51–2, 149, 153,
    155
eclipse **136**, 146, **147**
elements, four 38–9, 162
equinoxes 149, 158
Externsteine 16, **17**
Eye Goddess 155

Gallasus Oratory **14**
Glastonbury: Avalon 137; Chalice
    Well **9**, **139**, 140, **140**, 142,
    **142**, **143**, 143–4; Egg Stone
    141; Goddess Conference **144**,
    **145**, 145–6; Tor 126, **136**,
    139–40, 141, 143, **147**; White
    and Red Springs 127, 140,
    141, 144
Gotland, Sweden **85**
Grail 90, 139
Great Rite 124
Green Man 42, 134
Grimes Grave **16**
grounding 58
Guinevere 138, 139
Gundestrup Cauldron 16
Gwynn-ap-Nudd 141

Hag's Chair 7, 159–63
Hawaii 17
holed stones 41–3

Imbolg 125
Innerwee's Backache Chair **110**,
    110–13

Joseph of Arimathea 138

Kennet, River 130
Kilnaboy 111–12, **112**
Knossos 16, 90
Knowth passage tomb 153, **153**

Lady of the Lake 98, 139
Lammas 138, 163
ley lines 13, 132, 137, 140–1
Long Man of Wilmington 74–5
Loughcrew passage tomb

**156**,156–8
Lough Gur **22**, 23–7, **25**, **26**

Madron's Well **116**, 117–18, **118**
Maes Howe 79, **87**
Maeve's Seat **158**
Maid **141**; rites 73
mazes 90, **90**
Men-an-Tol 41, **41**
Merlin 30, 138, 141
mermaids: in Cornwall 88–9, **89**;
    in Orkney 81–4
Michael, Saint 140
Michael ley line 140
Midsummer 30, 57, 149
Midwinter 135, 149, 150, 153
mirror 88, 93–5
Morgan-le-Fay 139, 142
Morris dance 134
Mother rites 73–4

naming ceremony 40
Newgrange passage tomb 30,
    126, 153–5, **154–5**
nine, number 36
Nine Maidens 34–5, **34**
Notre-Dame de Roncier **98**

Orkney 79–83, 150

pathworking: elemental 44–9;
    the spring within the well
    106–9
Poulnabrone Dolmen **114**, 115

quartz 153–4

Ring of Broadger **78**, **80**, **82**, **95**
Ring of Kerry **10**
rite(s): cord 35–7; Crone 70–1;
    Great 124; Maid 73; Mother
    73–4; of passage 70–6;
    seasonal 124–5

St Bridgit's Well 99–100
St Colman's Well 103–5, 109
St Nectan's Kieve 90–2, **92**
Sainte-en-Tregastel **10**
Samhain 30
Selkies 84–6
Serpent Mound 16
sheela-na-gigs 42, 111–12, **113**,
    124
'ship' site **85**

Silbury Hill **122**, **127**, **164**;
    creation 126–7, 132; function
    123; Goddess image 127, 130,
    131
snakes 131, 132, 133, 140–1
spiral art **154**, 155, **162**
springs 97; at Callanish **68**
Steness, Stones of 79–80
Stonehenge: access 11, 31–3; ley
    line 140; observatory 30–1;
    sun-aligned 25; temple 31, **31,
    32–3**
Stones of Bro **81**
Stones of Riasc **29**
Stones of Stenness **82**
Swallowhead Spring 127, 130

Tintagel **91**
Tomb of the Eagles 150, **151**,
    151–2
tombs 149–50
toothache wells **102**, 115–16
Triple Form Goddess 34, 65–9,
    139, 158

Venus of Willendorf **16**

water: element in ritual and
    meditation 119–21; healing
    101; power of 101
wells: Bridgit's Well, **99**, 99–101,
    **100**; Colman's Well, 103–5,
    **108**; 'eye cure' **104**, **105**;
    healing meditation 100–1;
    horseshoe well **104**;
    pathworking 106–9; sacred
    sites 97–8, 102; Well of the
    Holy Cross **96**, White Cow Well
    **119**
West Kennet Long Barrow **166**;
    burial chamber 123, 150; visits
    11, 165, 166
White Cow Goddess 56
widdershins 93
Witchcraft Laws 19
Witches' tools **9**

Zennor 88
Züschen **43**